Fraud in the Lab

FRAUD IN THE LAB

The High Stakes of Scientific Research

NICOLAS CHEVASSUS-AU-LOUIS

Translated by Nicholas Elliott

 Harvard University Press

Cambridge, Massachusetts
London, England
2019

Originally published in French as *Malscience: De la fraude dans les labos* © Éditions du Seuil, 2016

Library of Congress Cataloging-in-Publication Data

Names: Chevassus-au-Louis, Nicolas, author.
Title: Fraud in the lab : the high stakes of scientific research / Nicolas Chevassus-au-Louis ; translated by Nicholas Elliott.
Other titles: Malscience. English.
Description: Cambridge, Massachusetts : Harvard University Press, 2019. | "Originally published in French as Malscience: de la fraude dans les Labos © Editions du Seuil, 2016."—Title page verso | Includes bibliographical references and index.
Identifiers: LCCN 2019006968 | ISBN 9780674979451 (alk. paper)
Subjects: LCSH: Fraud in science. | Research—Moral and ethical aspects.
Classification: LCC Q175.37 .C4413 2019 | DDC 507.2—dc23
LC record available at https://lccn.loc.gov/2019006968

Contents

Preface

In October 2014 Arturo Casadevall and Ferric C. Fang of the Albert Einstein College of Medicine published an amusing nosography of the sciences in the journal *Microbe*. Nosography is the branch of the medical arts that deals with the description and classification of diseases. To my knowledge, no scholars describe themselves as nosographers, which is a shame because I would have the greatest respect for those bold enough to devote their research to this delicate, subtle science perennially faced with two major challenges. The first is to gather some of the symptoms exhibited by the sick human body, using a well-thought-out classification, to constitute a disease. This has the unavoidable effect of creating sick people, since we all find it so comforting to know that what ails us has a name, particularly a scholarly one. The second is to forever be faced with the difficulty of distinguishing between the normal and the pathological, to borrow the title of a dissertation by Georges Canguilhem. This is a task not unlike sailing between Scylla and Charybdis: if you get too close to the normal, you run the risk of neglecting the experience of the sick person, who feels like a sick person, while if you stray too far into the pathological, you might turn into an overenthusiastic investigator, forcing any unique feature into the realm of disease. Casadevall and Fang's article reveals the existence of *amnesia originosa*, "an inability to recall the actual origin of an idea that one now regards as one's own." It also identifies *Nobelitis*, "a rare but debilitating condition afflicting only the scientific elite," which can manifest itself "by auditory hallucinations involving telephone callers with Swedish accents," generally in the fall (Nobel Prizes are announced by the Royal Swedish Academy of Sciences in the first two weeks of October), and most often turns into a depressive episode, unless the *Nobelitis* develops into *hyperpromotosis*, in which "the recurrent overestimation of the importance of one's findings and the zeal exhibited in broadcasting one's accomplishments are pathognomonic signs." The suggestion here is that some researchers are ungrateful, vain, and avid for recognition. In fact, the history of science is rich with examples of the type. But Casadevall and Fang also describe some new, previously unreported pathologies, including premature publication syndrome, affecting individuals who submit unfinished manuscripts to journals out of fear of being overtaken by the competition; *areproducibilia*, "the inability to obtain the same experimental result twice," a syndrome that is "not necessarily a problem for individuals who publish irreproducible results and simply move on to leave other scientists to deal with the problems"; and *impactitis*, an obsessive condition characterized by the conviction that the "value of scientific work is based on the impact factor of the journal where the work is published" (that is, the average number of article citations published from this journal in the previous two years) rather than the work's intrinsic qualities.

Anyone with the slightest experience of the world of laboratories and universities has probably heard of a few individuals afflicted with these pathologies. But for those who are foreign to the academic microcosm, this epidemiological snapshot might be worrisome. This book is addressed to both categories of readers, with the additional intention of showing that Casadevall and Fang's nosography is incomplete. The authors state that "there is no known cure" for *impactitis*, "a highly contagious and debilitating condition." Sadly, their conclusion is plausible. I aim to show that the disorder is not only apparently incurable but also associated with other formidable pathologies not described by the two nosographers: data beautification, which transforms an often inconsistent collection of observations into a superb scientific story with an impeccably worked-out narrative; plagiarism, an extreme form of *amnesia originosa*; and finally, plain and simple fabrication of scientific results, downright forgery, fraud, and invention.

Etymology invites my use of the metaphor of nosology. One of the possible roots of the word *malady* is *mal habitus,* or that which is in a bad state. But any reader of Pierre Bourdieu will also recognize *habitus* as the concept the sociologist used to refer to the system of embodied social dispositions that result in each person behaving exactly as expected in a given social environment, because the individual has internalized "a system of schemes of perception, thought, appreciation, and action" specific and suitable to this environment. My sense is that today a certain scientific habitus is ailing: the habitus by which research was an open world dedicated to the pleasures of the intellect, a world whose habits included casual first-name address, a relative indifference to social hierarchies, and a propensity for mockery and irreverence, and in which money was of secondary importance, all of which boldly stood out in comparison to the mores of other social environments requiring an equivalent degree of intellectual certification. For many years, a scientific researcher could be distinguished from an engineer, a lawyer, a doctor, or a financial analyst at a single glance. Researchers stood out by both their appearance and their speech. This was a part of their habitus.

Yet after spending time in laboratories as an observer over the last fifteen years and working in them as a researcher while writing my dissertation, I have to recognize that this habitus is in a bad way. To tell the truth, I can barely recognize these places I was once so fond of. Yes, they are (still) places where people talk about science, but they do so far less than they discuss publication strategies, funding, recruitment, competition, visibility, recognition, and, to borrow the local terminology, "principal investigators," "leading projects," and "first deliverables." The phrase "publish or perish" is not new—it is said that the physiologist Henri Laugier, appointed the first director of the Centre national de la recherche scientifique (CNRS) in 1939, displayed it on his office wall—but it has never so accurately described the workings of the world of science. The idea expressed in that short phrase has led to the scientific fraud and various types of *malscience* discussed in the following chapters.

One cannot collectively blame researchers for the developments that have turned the lab into a factory producing scientific knowledge, much of which is of dubious quality. Many of them proudly resist this new norm of quantitative evaluation, which punishes the nonpublished and showers praise on the overpublished. And there is no shortage of local initiatives on the part of laboratories, universities, and research centers attempting to put a little sand in the gears of this terrible machine.

These acts of resistance are undeniably laudable. They give us hope that we can confront the explosion of scientific fraud I describe in this book. Using numerous examples—some widely covered in the media, others still concealed in the hushed insiders' sphere of the scientific community—I have attempted to quantify scientific fraud, based on the proliferating number of studies showing the growing proportion of publications that are more or less falsified. I also endeavor to show the profoundly serious consequences for all those who rely on biomedical research to find a solution to their ills and blindly place their trust in researchers, beginning with those suffering from disease. Most importantly, I show that this explosion of fraud proceeds from the global reorganization of laboratories, which has introduced excessive competition, extreme individualism, and short-term incentives—in other words, the worst of the corporate world.

Some of those who commented on the manuscript criticized me for making the situation seem worse than it is and for painting a damning picture of a scientific community that, despite the growing number of cheats and frauds in its midst, remains collectively committed to the values of intellectual integrity and honesty. I understand this criticism but cannot agree with it.

They say you're toughest on the ones you love. In this case, it is not a question of being tough, but simply of describing. And, in any case, despite it all, of loving.

Fraud in the Lab

1

Big Fraud, Little Lies

The mathematician Charles Babbage is now considered one of the originators of the concept of the computer, but he was also a fierce critic of the English scientific institutions of his time. In his *Reflections on the Decline of Science in England*, Babbage devoted a few juicy pages to distinguishing between four categories of scientific fraud.[1]

The first is hoaxing. Babbage uses the example of an individual known as Gioeni, a knight of Malta who in 1788 published a meticulous description of a little shellfish. Gioeni claimed the honor of naming this shellfish—which he had also entirely invented. Babbage finds this practice difficult to justify, unless it is used to ridicule a scientific academy that has "reached the period of dotage." Advice worth remembering.

The second type of fraud is forging data, or simply making it up. As Babbage defines forging, it "differs from hoaxing, inasmuch as in the latter the deceit is intended to last for a time, and then be discovered, to the ridicule of those who have credited it; whereas the forger is one who, wishing to acquire a reputation for science, records observations which he has never made." He provides the example of the Chevalier D'Angos, a French astronomer who in 1784 described the passage of a comet that had never existed, having imagined his "observations" by drawing from various astronomical tables.

To describe the third type of fraud, Babbage borrows a metaphor from the art of gardening: this is the practice of trimming experimental data. "Trimming consists in clipping off little bits here and there from those observations which differ most in excess from the mean, . . . a species of 'equitable adjustment,' as a radical would term it." Though he was politically aligned with the radicals, Babbage made clear that this approach could not be accepted in science. However, he considered this kind of fraud relatively harmless in that the average of the results remains the same, given that the object of the manipulation is only "to gain a reputation for extreme accuracy in making observations."

The last category, about which Babbage has the most to say, is cheerily referred to as the cooking of data: "This is an art of various forms, the object of which is to give to ordinary observations the appearance and character of those of the highest degree of accuracy. One of its numerous processes is to make multitudes of observations, and out of these to select those only which agree, or very nearly agree. If a hundred observations are made, the cook must be very unlucky if he cannot pick out fifteen or twenty which will do for serving up."

First Scandals

Babbage's book indicates that scientific fraud in all its manifestations is hardly a novelty. The scholarly community long knew of its existence but never referred to it publicly. Things changed in the 1970s, when several scientific frauds hit the headlines and were openly discussed in the United States. Thanks to the mistrust of government following the Watergate scandal and the decade's general antiestablishment climate, cases that would previously have remained secret in the labs had a widespread impact. The following are some of the principal instances.

To my knowledge, William Summerlin was the first perpetrator of scientific fraud whose misconduct was covered by the mainstream press. In the early 1970s he was a young doctor doing research in immunology at the Memorial Sloan-Kettering Institute for Cancer Research in New York. To study tissue transplants, Summerlin was using skin grafts on mice as an experimental model. But it soon became apparent that the researcher had touched up tissue that had been implanted from black mice to white mice with black ink, to make it look as though the skin grafts had taken. Summerlin readily admitted to his misdeed, attributing it to nervous exhaustion and the pressure to obtain results. He was granted extenuating circumstances, with a year's forced leave and the obligation to seek treatment "in order to allow him to regain the calm and professional vigilance his functions require." Summerlin later withdrew to Louisiana to practice dermatology, and his legacy is of a mouse colorist. In one of the first scholarly works on scientific fraud, Marcel C. LaFollette observed that "[the] political framework adopted for post-1945 organization of U.S. science assumed that scientists were trustworthy and that their technical expertise was always reliable."[2] This trust began to collapse after the Summerlin affair.

The 1970s also witnessed the sinking of the posthumous reputation of one of the leading authorities of British psychology, Sir Cyril Burt. When he died at the age of eighty-eight in 1971, Burt left a respected body of work, particularly focused on analyzing the heritability of intelligence. By comparing pairs of monozygotic, or identical, twins (that is, twins having developed from the fertilization of a single egg and therefore sharing the same genetic makeup) and dizygotic, or fraternal, twins (that is, twins having developed from the independent fertilization of two eggs and therefore having the same genetic relatedness as a brother and sister), Burt established that a person's score on an IQ test was 75 percent heritable. But beginning in 1974, some skeptics noted that the IQ scores Burt used in study after study were always identical to a few hundredths of a unit. It was hard to prove that Burt had tampered with his results, given that he had destroyed his archives. Yet today there is no doubt that the psychologist got carried away by his passion for eugenics, which drove him to demonstrate what he wanted to prove and led him to invent most of his results.

Summerlin used the rustic tricks of a horse trader, and Burt let himself get wrapped up in his ideological passion. A few years later, Mark Spector took it upon himself to provide a new variation on the fraudulent researcher, one that would live on for decades: that of the ambitious young scientist eager to publish apparently sophisticated, cleverly doctored results in the most prestigious journals. In 1981 *Cell,* a highly respected journal of cell biology, published an article with Spector as first author. In it, Spector claimed to have discovered a new biochemical reaction pathway in tumoral transformation. The press saw Spector's claim as a tremendous advance in cancer treatment research. However, those who took a closer look at the details found themselves a little more suspicious. In a relatively insignificant figure illustrating the article in *Cell,* Spector presented results from a preparation of a protein that had been phosphorylated by means of this promising reaction, but another figure revealed that the purification of this protein required years of work.

How could a young doctoral student get all this work done while simultaneously publishing six articles describing different aspects of his alleged discovery? His PhD adviser, Efraïm Racker, had the integrity to heed these skeptical remarks. He studied his student's experiments closely and publicly recognized that he had been careless in supervising this doctoral candidate, who had cleverly fabricated the results of fake experiments. After Spector was expelled from the university, Racker retracted all of his former student's articles.

A few years later, the John R. Darsee scandal would cast doubt on the integrity of researchers at one of the most reputable institutions in the world: Harvard Medical School. Darsee was a workaholic, a young cardiologist considered one of the prestigious school's most promising students of the early 1980s. In May 1981 he had just finished a postdoctoral fellowship that would apparently lead to his being recruited by Harvard, when he was caught by his colleagues in the act of tampering with an experiment. He swore that this first misstep would be the last. His apologies were accepted. Six months later, new doubts arose when Darsee's work, which identified significant biochemical changes in patients with cardiac problems, proved impossible to reproduce. Both the National Institutes of Health (NIH) and Harvard Medical School launched investigations. It soon came to light that Darsee had invented or falsified his data before, during, and after his postdoc. In 1983 Harvard's faculty of medicine, which had meanwhile been forced to reimburse the NIH for grants misused by Darsee (the first time such a measure had been taken), demanded the retraction of about thirty articles by its former researcher. The incident stood out both for the level of fame of the institution tarnished and for the unprecedented number of fraudulent articles. In its June 9, 1983, issue, the prestigious and venerable Boston-based *New England Journal of Medicine* included two retraction notices for articles it had published by Darsee, a first since it was founded in 1812. The same issue featured a letter from Darsee in which he stated that he was "deeply sorry for allowing these inaccuracies and falsehoods" in the two articles published. He extended his apologies to the editorial board and readers of the *New England Journal of Medicine,* as well as to his coauthors, "impeccably honest researchers" who were not aware of his wrongdoing. This public apology was equally unprecedented, yet proportionate to the scandal.

Defining Fraud

As I'll have occasion to repeat, scientific fraud is hardly a novelty. While it is inherent to research, it became a subject of concern in the 1970s and 1980s, which saw the scientific community rocked by the four momentous cases just described. As Anne Fagot-Largeault, professor emeritus of the philosophy of medicine at the Collège de France, writes, "Until the 1980s, scientific fraud (or misconduct) mostly interested historians of science, who published entertaining, sometimes astounding case studies, as gripping as detective novels. . . . But beginning in the 1980s, lawyers, moralists, politicians, research administrators, and scientists themselves (reluctantly) started confronting the issue."[3] However, as Fagot-Largeault points out, they only confronted the issue with the conviction that fraud was the work of black sheep who could be separated from the herd. In 1983 the annual meeting of the American Association for the Advancement of Science, the prestigious scholarly association that publishes the journal *Science,* titled one of its sessions "Fraud and Secrecy: The Twin Perils of Science." This was in and of itself a remarkable event, given how taboo the question had been. Yet most of the speakers emphasized that fraud remained highly marginal and was perpetrated by researchers who lacked professional ethics and a moral sense and who were possibly even psychologically disturbed. As the president of the National Academy of Sciences testified before a congressional investigative committee in 1981, "One can only judge the rare acts that have come to light as psychopathic behavior originating in minds that made bad judgments—ethics aside—minds which in at least one regard may be considered deranged."[4] In other words, the scientific community agreed to consider the issue of fraud only to immediately deny its importance.

By recognizing that it was being undermined by fraud, the scientific community was faced with the question of defining fraud. Little by little, two definitions emerged. The first is American. In 2000 the US Office of Science and Technology Policy defined the breach of scientific integrity as the fabrication, falsification, and plagiarism of data (summed up by the acronym FFP), whether in the conception of research projects (particularly in the writing of grant proposals), their execution and publication, or the reviewing of articles by referees. The FFP definition of breaches of scientific integrity only applies to manipulations of experimental data. It does not deal with professional ethics. The second definition, which is often used in Europe, expands the scope of breaches of scientific integrity by including what international institutions such as the Organisation for Economic Co-operation and Development often qualify as "questionable research practices": conflicts of interest, a selective and biased choice of data presented in articles, "ghost" authors added to a publication in which they did not participate, refusal to communicate raw experimental data to colleagues who request it, exploitation of technical personnel whose contributions are sometimes not recognized in publications, psychological harassment of students or subordinates, and failure to follow regulations on animal testing and clinical trials.

While I adhere to the European conception of scientific misconduct, breaches of scientific integrity are increasingly frequent no matter how we define them. Yesterday's alleged exceptions are now becoming the norm. A growing number of articles and editorials in scientific journals express outrage at the decline of scientific rigor in laboratories. As Jean-Pierre Alix, a former manager at the Centre national de la recherche scientifique (CNRS), puts it, "We're in the process of digging up the corpse of scientific fraud."

Some Disciplines Are Spared

One should make clear from the outset that not every discipline is equally concerned. While the field of mathematics has seen some cases of plagiarism, it is free of the problems of data fabrication and falsification: a result is either correct or incorrect, and an editorial board's discussion only focuses on the value of publishing the new result. In physics, there is a marked tendency to publish, or at least communicate, results too quickly, but the scientific community in question tends to effectively self-regulate. Admittedly, the European

Organization for Nuclear Research did prematurely announce in 2011 that it had detected neutrinos moving slightly faster than the speed of light, a speed that current theories consider unsurpassable, but a few months later it recognized that the observation had been skewed by technical problems with the measurement devices in its particle accelerators. Similarly, American astrophysicists with the Bicep-2 program were certainly a little hasty in stating in the spring of 2014 that, for the first time, a radio telescope in Antarctica had allowed them to detect gravitational waves that would be the first experimental argument supporting the theory that our universe was created by a Big Bang. By fall, a rival team, the Planck collaboration, had demonstrated that the signal detected could show something other than the gravitational waves in question.

Overall, physics has been spared the epidemic. Why? The theoretical understanding of reality at the core of physicists' work is more solid, while their experiments are longer and more soundly connected to experimental hypotheses. All the physicists to whom I have spoken believe that fraudulent articles pose no threat to their field. "I've never been faced with cases of obvious fraud, like data fabrication or even beautification," Pascal Degiovanni, a physicist at the École normale supérieure in Lyon, told me. "What can happen, however, is that people publish results based on limited experiments, without exploring everything that should be explored." Degiovanni does have real concerns about the future: "What is certain is that the more people are under pressure to obtain funding, the more probable it becomes that a team has to decide between quality and speed." Physicist Michèle Leduc, who was also the chairwoman of the CNRS Ethics Committee, agrees:

> There is no comparison between the rigor applied to collecting and treating data in physics and that in biology, which has protected our discipline from the difficulties currently encountered

by biology. As soon as an experiment is published, rival teams try to reproduce it. If they aren't able to, they publish a commentary in the specialized literature that will start a discussion to understand the reasons why. One can certainly note that more and more commentaries are being published, which suggests that physicists have a tendency to start publishing too fast, but unlike in biology, there is no increase in the number of retractions of fraudulent articles.

Along with these epistemological reasons, one must take into account the sociology of the sciences. While there is competition in physics, it generally applies to a handful of groups of people throughout the world, most of which consist of a few hundred researchers. Within these large teams, self-monitoring is not always immediate, but it is certainly effective. In 1999, for instance, Victor Ninov, a researcher at the prestigious Lawrence Berkeley National Laboratory, announced in *Physical Review Letters* that he had identified unknown atoms with an exceptionally high molecular mass. Elements 116 (livermorium) and 118 (oganesson), allegedly identified by Ninov, soon elicited perplexed reactions from specialists in the field. Ninov's many coauthors took another look at the experimental results and quickly realized he was the only one who had been in a position to arrange the data. Three years after the initial publication, *Physical Review Letters* published a corrigendum pointing out all Ninov's voluntary errors in analyzing the experimental data. In the meantime, he had been fired by the Lawrence Berkeley National Laboratory.

On the other hand, research in chemistry, biology, and especially the social sciences is the work of countless very small groups, sometimes only consisting of a single individual with a few students and technicians. Consequently, peer monitoring is less vigilant on a daily basis. Whether fraud is due to an ambitious young researcher, a student in a rush to stand out, or a confirmed team leader, there will be no more than a handful of colleagues to verify the results with a critical eye. This is most often not the case in physics, where articles are attributed to hundreds of researchers, or even collectively attributed to an institute or a team, which from a mere statistical perspective makes it unlikely that no one will stand up to denounce any potential misconduct before publication.

The Explosion of Retractions in Biology

One of the clearest signs of the proliferation of breaches of scientific conduct is the explosion in the number of article retractions. Two researchers, Michael Grieneisen and Minghua Zhang, took it upon themselves to survey a corpus of over four thousand articles retracted from all fields of scientific literature since 1928.[5] Their first conclusion was that retractions are a recent phenomenon: they only found twenty-one before 1980. Their second conclusion was worrisome: the proportion of retracted articles is ballooning, increasing elevenfold between 2001 and 2010. And this figure was obtained by excluding serial fraudsters like Fujii, Schön, and Boldt, discussed in Chapter 2. The third conclusion is reassuring: the retracted articles only account for a minute portion of the scientific literature. Still excluding the retraction of dozens of articles by serial fraudsters, one finds a relatively low rate of retractions out of the total number of articles published: 0.16 percent in *Cell,* 0.09 percent in *Science,* 0.08 percent in *Proceedings of the National Academy of Sciences,* 0.05 percent in *Nature.* These journals are both the most prestigious and those that have the highest number of retractions. In any case, these figures are very low, less than one article out of one thousand, but the strong increase in the rate of retractions over a decade remains troubling. The last lesson gleaned from the survey is that the medical field is the one most affected by the retractions epidemic. Publications in the field account for 40 percent of the corpus studied, but 52 percent of retractions.

However, Grieneisen and Zhang's study did not directly focus on the causes of retractions, which can also be made in good faith, when scientists discover a past error and notify their colleagues. This is a difficult distinction to make. There is nothing to prevent researchers from claiming they made a good-faith error rather than running the risk of appearing to have perpetrated fraud. One cannot deny that certain retractions are the product of genuine intellectual honesty. Experimental research in biology and medicine is subject to count-less contingencies, which I will discuss at greater length in Chapter 5. Reagents can be poorly prepared, samples mixed up, test tubes in-correctly labeled. In an environment where lab personnel change from month to month based on the comings and goings of students and postdocs, collaborators can be discovered to lack rigor or even be truly careless after the fact. Error is always possible. And it is to the credit of eminent scientists such as Linda Buck, winner of the 2004 Nobel Prize in Physiology or Medicine, and the plant gene-ticist Pamela Ronald, that they retracted some of their publications after realizing they were based on erroneous experimental data.

What is the proportion of fraud in these increasingly frequent re-tractions? The American psychiatrist R. Grant Steen, owner of the consulting firm Medical Communications Consultants, was the first to dive into the murky waters of articles officially retracted from the biomedical literature. In a study published in 2011, he analyzed a body of 742 English-language articles published in the first decade of the twenty-first century.[6] The question he asked himself was a simple one: Were these articles withdrawn following the discovery of a good-faith error or of fraud? At first glance, his study seems to re-inforce an idea deeply rooted among researchers, which is that retrac-tion is a useful self-correction mechanism that allows scientists to publicly recognize their errors, which are inherent to the profession, ensuring that their colleagues do not base their own work on erroneous findings. Fraud, when defined as fabrication or falsification of experimental data, makes up only 26 percent of the retracted articles in the corpus studied. Three-quarters of the articles were therefore retracted for "scientific error."

But the description of the error is often vague. In fact, it is simply not described in 18 percent of cases. Most significantly, Steen notes, the rate of retractions grew between 2000 and 2010. "It seems impossible that both fraud and error could both increase so quickly and in such a correlated manner," the author states. Between 2004 and 2009, the number of articles retracted for fraud increased sevenfold and those retracted for error doubled, while the total number of articles published increased by only 35 percent. Steen suggests that "a possible interpretation of these results is that the incidence of fraud in research is increasing."

This suspicion was spectacularly confirmed by the publication in 2012 of a study by two researchers working with Steen at the Albert Einstein College of Medicine in New York.[7] The authors had examined the retracted articles in the Medline database, which lists and indexes nearly every scientific article published internationally in the life sciences and medicine. According to their calculations, only 21 percent of the 2,047 articles retracted since 1973 were withdrawn after a good-faith error was identified. The leading cause of retraction, by far, is fraud, confirmed or suspected: it accounts for 43 percent of retractions (Fig. 1.1). The other reasons are duplicate publication in 14 percent of cases (standard practice is to publish the results of a study only once) and plagiarism of other articles in 9 percent of cases; the rest are due to conflicts between authors. As the study emphasizes, the rate of retraction for fraud has constantly increased: it has grown tenfold since 1975 and has now reached 0.02 percent.

The epidemic is particularly intense among the most prestigious journals, which are those whose articles are the most cited. Aside

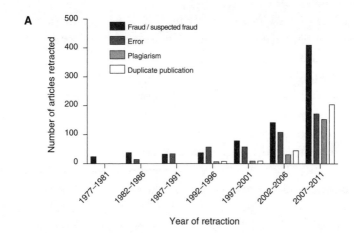

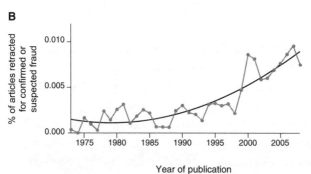

Figure 1.1. Retractions in the biomedical literature have skyrocketed since the early 2000s. The four causes of retraction (fraud or suspected fraud, good-faith error, plagiarism, and duplicate publication) are all increasing (A), but retractions for suspected or confirmed fraud have increased the most (B). (Source: F. C. Fang, R. Grant Steen, and A. Casadevall, "Misconduct accounts for the majority of retracted scientific publications," *Proceedings of the National Academy of Sciences* 109, no. 42 (2012), fig. 1.)

from *Acta Crystallographica Section E,* which is only in the lead because it retracted no fewer than seventy articles by a pair of fraudulent Chinese researchers in 2009, the most retractions have been published by *Science* (thirty-two articles), *Nature* (nineteen articles), and *Cell* (twelve articles), all journals in which every biologist dreams of publishing. One even finds a narrow correlation between a journal's impact factor (that is, the average number of citations of papers published in the journal in the previous two years) and its rate of retraction for fraud and error. There is another sign that this problem has struck the very heart of the global scientific system: researchers in the United States, Germany, and Japan, three nations with long-established, first-rate scientific reputations, are responsible for two-thirds of the articles recognized as fraudulent. As for France, it does not even appear in the article, which may be a sign that its contribution to biomedical research has become negligible on an international scale. Unless the national culture makes French researchers more reluctant to acknowledge their faults? I'll return to this no less disturbing theory in Chapter 14.

Even more worrisome is that we have serious cause to think that there is far more fraud, or more generally, scientific misconduct, than bibliometrics can identify. Several studies on the sociology of scientific knowledge sent out anonymous questionnaires to thousands of researchers in an attempt to quantify the everyday breaches of scientific integrity that could eventually lead to actual fraud. Daniele Fanelli, a researcher in the sociology of scientific knowledge then at the University of Edinburgh, synthesized twenty such studies carried out between 1986 and 2005.[8] His conclusions showed that 2 percent of scientists acknowledge having fabricated or falsified results at least once in their career—a percentage one hundred times greater than the rate of retraction of articles for fraud. Additionally, there are significant reasons to believe that this rate is underestimated, as is the case any time respondents are asked to declare whether they have broken ethical rules, whether or not the survey is anonymous.

In fact, 14 percent of scientists report that they are aware of colleagues who commit fraud—in other words, seven times more than those who acknowledge having committed fraud at least once. Once we move away from the deadly sin of fraud and enter the gray zone of venial sins (such as using a method of analysis known to no longer be the most pertinent but liable to yield a desired result, changing methodology in the course of a study, or excluding from an article those experiments that do not confirm its thesis), the rate of acknowledgment leaps to 34 percent. Fanelli's research covers studies carried out in all the disciplines, but once again, the meta-analysis shows that research in the fields of biology and medicine is overrepresented.

Let's recap. According to the biometrics of retracted articles, one out of five thousand articles in biomedicine is fraudulent. But if one is to believe the sociological surveys, only one in one hundred cases of scientific misconduct is identified. This suggests that the rate of fraudulent articles is in the range not of 0.02 percent but of 2 percent. Unfortunately, this is not the only reason to think that the scientific literature is deeply tainted, particularly in biomedicine.

2

Serial Cheaters

On August 5, 2014, the Japanese biologist Yoshiki Sasai hanged himself in the offices of the Riken Institute's Center for Developmental Biology in Kobe, a research center for which he served as associate director. While the content of several letters found among his belongings has not been made public, it is known that at least one letter was addressed to Haruko Obokata, a young researcher whose work Sasai supervised. Eight months earlier, Obokata had been the first author of two articles published in the prestigious scientific journal *Nature*.

The experimental results described in these articles seemed breathtaking. For over fifteen years, biologists around the world had been fascinated with stem cells, a type of cell that can both divide indefinitely and differentiate into any kind of cell found in the human body. The ability to culture stem cells would enable a form of regenerative medicine in which tissues damaged by disease would be replaced by these therapeutic cells. Unfortunately, the isolation and culture of stem cells remains complex, and control of their differentiation is still rudimentary. Yet in the January 30, 2014, issue of *Nature,* thirty-two-year-old Obokata and her thirteen coauthors announced that they had discovered a disarmingly simple method of transforming an adult lymphocyte (a type of white blood cell) into a pluripotent stem cell—in other words, a cell capable of differentiating into countless types of cells. According to them, one merely needed to immerse the lymphocytes in a slightly acidic solution for half an hour. Once injected into a mouse, the so-called STAP cells (for *stimulus-triggered acquisition of pluripotency*) thus obtained proved capable of differentiating into any type of cell, including that of a placenta, a result never previously observed with stem cells. Better yet, the Obokata method displayed a yield thirty times greater than that of the best previously known methods for obtaining pluripotent stem cells.

Dozens of laboratories throughout the world immediately attempted to produce these miraculous STAP cells. All failed. Having gotten their fingers burned, researchers grew suspicious. What if Obokata had committed research fraud? What if she had invented or beautified her data? The administration of the Riken Institute took the rumors seriously and launched an internal investigation. Early in April 2014 its findings were announced. They were damning.

In manipulating the image data of two different gels and using data from two different experiments, Dr. Obokata acted in a manner that can by no means be permitted. . . . Given the poor quality of her laboratory notes it has become clearly evident that it will be extremely difficult for anyone else to accurately trace or understand her experiments . . . . Dr. Obokata's actions and sloppy data management lead us to the conclusion that she sorely lacks, not only a sense of research ethics, but also integrity and humility as a scientific researcher. We were also forced to conclude that the normal system by which senior researchers should have been carefully checking all raw data did not work in this case. . . . Drs. Wakayama and Sasai allowed the papers to be submitted to *Nature* without verifying the accuracy of the data, and they bear heavy responsibility for the research misconduct that resulted from this failure on their part.[1]

On July 2, 2014, under extreme pressure from the editors of *Nature*, Obokata and her collaborators decided to request to retract their articles, which amounts to erasing them from the scientific literature.

Barely a month later, Sasai ended his life. The Riken Institute's investigation committee had emphasized that he was in no way complicit in Obokata's fraud and had only criticized him for falling severely short in his supervision of her work. But Sasai stated he was overwhelmed with shame. Charles Valensi, the other experienced scientist involved in the brief saga of the STAP cells, notified his colleagues at Harvard that he intended to take a sabbatical. As for the editors of *Nature*, they were deeply embarrassed by the online publication of peer-review reports by the experts—or "referees"—who had read Obokata's manuscripts and pointed out their deficiencies. Why did the prestigious British publication choose to ignore these criticisms and publish work that specialists found suspicious?

Korean Cloner

According to Karl Marx, history repeats itself: the first time as tragedy, the second as farce. Marx's observation remains accurate in describing scientific fraud, but the order has been reversed. While Sasai's suicide put a tragic end to a case that was the talk of the little world of stem cell biology for a semester, it had been preceded a decade earlier by a comparable case whose outcome was practically laughable.

In February 2004 the South Korean biologist and veterinarian Woo-Suk Hwang published two articles in *Science* announcing spectacular discoveries in the field of human therapeutic cloning. The researcher's team claimed to be the first to have cloned a human embryo to obtain lineages of stem cells indispensable for regenerative medicine. These results made international headlines, and Hwang became a celebrity in his homeland. It was expected that he would be the first Nobel laureate in Korean history. The Korean national airline even offered him lifetime free passage. In 2005 Hwang treated himself to another big splash, returning to the pages of *Science* to publish a description of the first cloning of a dog. Though the birth of the dog Snuppy did not make the international splash that the birth of Dolly the sheep did in 1997, it was nonetheless recognized as a significant advance in the biotechnologies of cloning.

But, as with Obokata and Sasai, the bloom was soon off the rose. In this case, the problem was not that the results were impossible to reproduce. In a field as complex as cloning, particularly one that is under strict legal restrictions in several countries—South Korea, however, is notoriously lax—few researchers attempted to reproduce Hwang's experiment. His initial problem appeared on a front he had neglected: ethics. One of his American collaborators accused him of not revealing his research objectives to the young women from whom he had taken the ova required for his cloning work. This could pass as a minor sin. But now the cloud of suspicion formed. As focus turned to Hwang's publications and scholars began dissecting his graphs and charts, it became increasingly clear that he had committed fraud. In December 2005 Hwang was forced to recognize his misconduct—he had retouched photographs and faked results—and his articles on human cloning were retracted from *Science*. In the wake of these revelations, the scientist was fired from Seoul National University and was sentenced to two years in prison for fraud, embezzlement, and violating bioethics laws. The ruling was reduced to a six-month suspended sentence on appeal.

Are offenders always punished? Only if we ignore Hwang's astonishing comeback and his ability to redirect his fate from tragedy to farce. While his articles on human cloning have been retracted, the article on dog cloning remains in the scientific literature. In 2006 Hwang made the most of this credit by founding the Sooam Biotech Research Foundation. The purpose of this allegedly nonprofit foundation is to clone household pets for the modest sum of

$100,000 per animal. When prospective clients did not beat down its door, the Sooam Biotech Research Foundation had the clever idea to organize a contest in the United Kingdom whose lucky winner would get to have his or her favorite animal cloned. While the selection criteria remain mysterious, in April 2014, Rebecca Smith, a young Londoner who deeply loved her aging female dachshund, Winnie, celebrated the birth of mini-Winnie, a clone produced by Hwang. After a brief eclipse in the late 2000s, Hwang returned to publishing the findings of his research on cloning as if nothing had ever happened, getting back to his solid prescandal rate of one article every two months.

Dutch Psychologist

Sasai, Obokata, and Hwang were working in developmental and stem cell biology, a highly competitive field that can potentially yield significant financial benefits. One might attribute their misconduct to the lure of money and the desire to break through the global competition. But massive fraud has also occurred in far less competitive disciplines. Social psychology is considered neither a lucrative science nor a fashionable one. It is not known for attracting international headlines. Yet in 2011, an eminent Dutch specialist in this branch of psychology, Diederik Stapel of Tilburg University (Netherlands), was made to retract 55 of his 130 articles. His research focused on the genesis of social stereotypes. For instance, one of his studies published in *Science* and since retracted demonstrated that racial stereotypes become more frequent with an increase in economic insecurity.

An investigation conducted by Tilburg University revealed Stapel's method. Stapel would devise the plan for the experiment with his colleagues and numerous students and develop questionnaires aimed at pinpointing specific stereotypes. He would then tell his collaborators that the study, which required hundreds of people to respond to the questionnaire, would be conducted by his contacts among re-search assistants at other universities. In fact, Stapel did not ask for any questionnaires to be filled out. Instead, he personally generated the data that conveniently proved the hypothesis his research was supposed to test. While Sasai, Obokata, and Hwang falsified their data, Stapel simply fabricated it. It remains surprising that he was so easily able to dupe his colleagues, and for such a long time. The investigation committee at Tilburg University emphasized this in its report: "[Stapel's] data and findings were, in many respects, too good to be true. The research hypotheses were almost always confirmed. The effects were improbably large. . . . It is almost inconceivable that coauthors who analyzed the data intensively, or reviewers of the in-ternational 'leading journals,' who are deemed to be experts in their field, could have failed to see that a reported experiment would have been almost infeasible in practice, [and] did not notice the reporting of impossible statistical results."[2]

In fact, the alarm that set off the internal investigation was sounded not by international competitors but by three brave young re-searchers at Tilburg University. Stapel quickly admitted that he had invented most of the data in his articles of the last decade. After being fired by the university, he struck a deal with the prosecutor and was sentenced to 120 hours of community service for tax evasion, given that the money he received to fund his research was spent without any actual research being conducted under his supervision. Since his downfall, Stapel has expressed his regrets and profusely apologized: "I couldn't deal with the pressure to publish. I wanted too much, too fast. In a system where there aren't a lot of checks, where you work alone, I took the wrong direction." In his autobiography, *Faking Science: A True Story of Scientific Fraud,* he again rued his excessive taste for "digging, discovering, testing, publishing, scoring, and applause":[3] six words that perfectly capture the downward spiral into scientific fraud. The latest news is that the fallen researcher has embarked on a new career as a philosophy professor at the

Fontys Academy for Creative Industries, also in Tilburg. This private school offers to teach prospective students the art of creating "superb concepts, exciting concepts, moving concepts, concepts that improve quality of life." No doubt the 2,500 students enrolled will benefit from studying this difficult art with Professor Stapel.

American Neuroscientist

The fall of Stapel was contemporary with that of another well-known psychologist, Marc Hauser, a Harvard professor forced to resign in 2011. While his known fraud is less extensive than Stapel's (only one of his articles, published in the journal *Cognition,* has been retracted), it drew a storm of attention because it took place at one of the most prestigious scientific institutions in the world. Hauser had taught at Harvard since 1998 and specialized in animal and human cognition, straddling the line between biology and psychology. He had several times been elected "most popular professor of the year" by the Harvard student body. His research dealt with popular subjects such as monkeys' ability to recognize themselves in mirrors and to learn conceptual and abstract rules. His book *Moral Minds* was a best seller, praised by the linguist Noam Chomsky. Brilliant on both the scientific and media fronts, Hauser was one of the university's stars.

This promising career hit its first speed bump in the summer of 2010, when the *Boston Globe* announced that an investigation had been opened into "scientific misconduct" on the part of Hauser. At the same time, Hauser's 2002 *Cognition* article was retracted. Harvard's administration could only confirm the information, stating that Hauser had been recognized as "solely responsible, after a thorough investigation by a faculty member investigating committee, for eight instances of scientific misconduct" involving problems with "data acquisition, data analysis, data retention, and the reporting of research methodologies and results." When classes resumed in the fall, Hauser called in sick and the rumors and questions began to spread.

In September 2012 the Office of Research Integrity (ORI), the branch of the NIH (see Chapter 13) that funded Hauser's research, released its findings. The report was damning for Hauser: six instances of fraud were identified in his work from 2002 to 2011. One of the most striking involved the retracted article in *Cognition.* This study stated that cotton-top tamarins *(Saguinus oedipus)* had been able to learn abstract rules through sequential syllables. The results supporting this conclusion were synthesized on a bar graph showing responses obtained with fourteen monkeys. The ORI report revealed that half this data was fabricated. In fact, a careful reading of the graph suggests the experiment was carried out with sixteen monkeys, while the article's methodology section only refers to fourteen. In another article, published in 2007 in *Proceedings of the Royal Society B,* Hauser and his collaborators claimed to show that rhesus monkeys understand human intention when the experimenter points a finger at an object. While the article states that thirty-one (of forty) monkeys allegedly behaved this way, only twenty-seven monkeys actually did. The other instances found by the ORI are in the same vein: false statement of experimental data, aiming to make it particularly spectacular.

Hauser's reaction is probably the most surprising thing about the whole case. According to the ORI report, he "neither admits nor denies committing research misconduct but accepts ORI has found evidence of research misconduct." Subsequently, a compromise was reached by which Hauser agreed to be supervised by a third party for any research conducted with federal funding over the following three years. But it appears Hauser will never have to submit to this obligation, because he put an end to his academic career.

German Physician

You may have noted that these spectacular cases of fraud affect fields—biology and social psychology—that are still young, with tenuous results and underdeveloped theories. You may even imagine that physics, as the hardest of hard sciences, could not fall prey to such transgressions. I must correct that impression by telling you the story of Jan Hendrik Schön.

Born in 1970, this German researcher was a specialist in condensed matter physics. After receiving his PhD from the University of Konstanz, he was immediately recruited by the prestigious American company Bell Labs. At Bell Labs, Schön worked on replacing silicon in superconductors with organic crystalline materials. He initially appeared to obtain extraordinary results, showing that these materials could be substituted for silicon to make transistors. Before long, *Nature* and *Science* were vying for his numerous articles—up to seven a month! But it was soon brought to the physicist's attention that some of the graphs he published appeared oddly identical from one article to the next. Schön claimed this was merely due to his absent-mindedness while preparing his manuscripts.

Few were convinced by his defense. In September 2002 Bell Labs set up a committee to investigate Schön and his collaborators. While the latter cooperated with the investigators, Schön provided increasingly outlandish explanations, such as his accidentally erasing crucial data or losing the samples used. The truth soon came out. Like the psychologist Stapel, Schön was fabricating his results, though his method was a touch more sophisticated: he used mathematical functions to generate plausible results to experiments that he never conducted. At this time, seventeen of Schön's articles, including all those published in the prestigious journals, have been retracted. Little has been heard from him since he had his doctoral degree revoked and was sanctioned with an eight-year ban on applying for research funding in Germany, where he has relocated. It appears that he is now working in the private sector.

However, it remains a mystery how Schön was singlehandedly able to deceive his numerous coauthors over a period of more than five years, given that the investigation committee concluded that he was solely responsible. Science journalist Eugenie Samuel Reich attempted to answer this question in her remarkable book *Plastic Fantastic*.[4] Her subtle analysis reconciles psychological and sociological factors, Schön's disposition, and the setting he worked in.

Founded in 1925 and priding itself on having employed close to a dozen Nobel laureates, Bell Labs was in the midst of an identity crisis when it recruited Schön. The American telecommunications company AT&T, of which Bell Labs was the research and development arm, had recently lost its monopoly in the United States. Deregulation and the opening to competition had left AT&T's management struggling for the company's survival (it would eventually fold in 2005). The parent company's financial difficulties led to increased pressure on Bell Labs. While researchers had previously been free to explore the most unlikely lines of research in an atmosphere comparable to that of an academic laboratory, the administration now asked them to privilege research liable to quickly lead to patent applications. Schön's work was exactly the kind of research that could rapidly give AT&T a significant technological advantage by allowing it to launch a new generation of transistors. His supervisors consequently avoided looking at his work too closely. As Reich points out, when Schön wrote an article, their comments were more likely to suggest rewordings to get around potential objections from peer reviewers than to examine it in a critical manner. For his part, Schön is a people-pleaser, "someone with a knack for figuring out what his supervisors wanted to see and then delivering it to them."[5] Schön was pleasant and polite—in other words, diametrically opposed to glory-seeking Narcissuses such as Obokata, Hwang, and Stapel. Reich claims that from his student years, Schön had gotten in the habit of fudging his data so it fit perfectly with the state of knowledge in the scientific literature, or with his colleagues' favorite hypotheses.

Schön's success could therefore be explained by the encounter between a conforming people-pleaser and an institution demanding results both striking and liable to have practical applications. But Reich's book also extensively considers the responsibility of a third party: the journals *Science* and *Nature,* which accepted seventeen of Schön's articles in four years. Wrapped up in the competition that has driven them for decades, the two periodicals fought over articles by this wunderkind of physics, made even more valuable by the fact that he worked for one of the most renowned institutions in the field. The articles were published in record time after being accepted. More troubling is that comments by the journals' peer reviewers were rarely taken into account. While these were generally favorable, they asked for extensive clarifications. As Reich points out, "The technical details the reviewers asked Schön to include would, if provided, almost certainly have made it easier for other scientists to follow up his work in their own laboratories," and thus to discover his deception.[6] The journals' editors chose not to pass on these requests to Schön, however, and published his manuscripts without these clarifications. *Science* even published one of Schön's articles after receiving only a single favorable recommendation, while standard practice is to wait for several peer reviews before deciding on publication.

French Imposters

While the Schön affair was coming to light, a scandal erupted in another branch of physics, confirming that the system of critical peer reviews central to the operation of scientific journals is often inadequate. It involved the outrageous twin brothers Igor and Grichka Bogdanoff, who are more often sighted on the sets of talk shows and in high-society dinners than in the lab.

Let's not dwell on the twins' doctorates in physics, awarded by the University of Burgundy in 1999 and 2002, after ten years of alleged toil (in other words, three years over the average amount of time required to complete a doctorate). It is common knowledge that so long as you have written a research thesis—and preferably a thick one—you merely need to wear down a PhD adviser's patience to obtain the title of doctor. Indeed, it is practically unheard of in the history of the French academy for an applicant who has defended a doctoral thesis to be denied the title. The jury's room to maneuver is limited to whether it will confer the doctorate with honors. On the other hand, few doctoral theses have been pronounced "a rambling jumble of paragraphs full of vagueness, confusion, and lack of understanding," which is how a group of experts from the CNRS described the Bogdanoffs' theses in a 2010 report. The report continues, "While suspicions arise upon the first reading, it took our experts a considerable amount of time to demonstrate and provide supporting arguments for this work's lack of coherence. Rarely have we seen such empty research disguised with such sophistication."[7]

Nonetheless, six journals, two of them highly reputed (*Annals of Physics* and *Classical and Quantum Gravity*), accepted articles by the Bogdanoff brothers in the early 2000s. How could journal reviewers and editors have been so oblivious as to agree to publish such ineptly researched articles? Admittedly, the Bogdanoffs' articles were more like shams than outright frauds, to the point that many specialists thought they were hoaxes. In November 2002 *Classical and Quantum Gravity*'s editorial committee felt obliged to release a communiqué stating that the Bogdanoffs' article should never have been published and that there had been an error in the peer-review process. Frank Wilczek, who later won a Nobel Prize in Physics, made a similar argument when he took over the editorship of *Annals of Physics* in September 2001, expressing his regrets that his predecessor had been somewhat neglectful in his editorial monitoring of the journal. Nonetheless, the Bogdanoffs' articles are still part of the scientific literature: the journals in question gave up on retracting them, perhaps out of fear of seeing the case taken to court, for the twins are notoriously prone to suing for libel if anyone dares to contest their scientific talents. The journals' editors did declare they were open to publishing criticism of the twins' articles. But do any physicists really want to waste their time refuting such nonsense?

International Record Holders

Yet the frauds of Obokata, Hwang, Hauser, Stapel, and Schön and the Bogdanoff brothers' mystifications look like fine craftsmanship compared with the deceptions of Yoshitaka Fujii and Joachim Boldt, both of whom brought the art of fraud into the industrial age. Fujii is Japanese, Boldt is German. Both are anesthesiologists who currently are the top two record holders for serial fraud, with a respective 183 and 88 articles retracted for data fabrication between 2011 and 2013. In both cases, the researchers simply invented numerous patients allegedly enrolled in the clinical trials described in their articles.

The case of Fujii, the reigning record holder for scientific fraud, is worth describing in some detail because it demonstrates how a scientific community handles suspicions that a scientist has committed fraud. In a word: by sweeping them under the carpet. By repressing them. By looking the other way, as in the worst family scandals. The first doubts regarding the quality of the work of this Japanese anesthesiologist, who specialized in the treatment of postoperative nausea, surfaced in 2000, when Peter Kranke and his colleagues at the University of Würzburg (Germany) printed an unequivocally titled letter in one of the journals that had published Fujii's work: "Reported data on granisetron and postoperative nausea and vomiting by Fujii *et al.* are incredibly nice!"[8] The title alone tells anyone familiar with the muted ways of scientific literature that Fujii is being accused of fraud. And indeed, the letter goes on to use statistical analyses to demonstrate clearly that the data in Fujii's articles could only have been invented.

Nonetheless, the editors of *Anesthesia and Analgesia* apparently did not heed the severe warning. They settled for running the letter with a response from Fujii, who appeared to play for time by making one placatory statement after another. A year later, Kranke and his collaborators reiterated their accusations by reanalyzing all the data published on the effectiveness of granisetron (a pharmaceutical traditionally used to prevent vomiting caused by anticancer treatments) in the treatment of postoperative nausea. They clearly demonstrated that Fujii, whose abundant publications make up two-thirds of the literature on the subject, was the only one to find the substance effective.[9] Yet this new warning initially had as little impact as the previous one. No investigation was started, no article retracted. The Fujii problem apparently came to the anesthesiology community's attention shortly thereafter, however. This is certainly one way of explaining why after the second warning, Fujii practically stopped publishing his work in anesthesiology journals, turning instead to a wide variety of medical journals specializing in gynecology, ophthalmology, surgery, and pharmacology. A code of silence seemed to have settled over the case. Everyone knew the Japanese doctor's data was fake, but no one dared to say so. When an international colloquium gathered in 2002 to release the first recommendations for postoperative nausea, a tacit agreement was made to avoid citing a single one of the seventy articles covering Fujii's no fewer than 7,200 (alleged) patients. As Martin Tramèr, the editor in chief of the *European Journal of Anaesthesiology,* observed in a remarkably frank editorial explaining the retraction of twelve articles Fujii had published in his journal, "Nobody, neither editors, nor peer reviewers nor readers, nor the manufacturer of granisetron, or even Fujii himself, asked why we had so overtly ignored his articles. Fujii had become invisible."[10]

The irony is that Fujii precipitated his own downfall by trying to put an end to his sudden invisibility in the anesthesiology community. In 2011, having not published in international journals specializing in anesthesia in four years, whether because of his own caution or implicit censorship on the part of editors, Fujii submitted an article to the *Canadian Journal of Anesthesia*. This time, the editors of the journal served as whistleblowers by sharing their suspicions with the institution for which Fujii worked, the Toho University Faculty of Medicine. An internal investigation was launched and quickly concluded that there had been fraud. Fujii was fired, and journals began retracting his articles in quick succession. Of the nearly two hundred articles he had published in twenty years, only nine are still in the scientific literature. But for how long?

3

Storytelling and Beautification

Is every scientific article a fraud? This question may seem puzzling to those outside the scientific community. After all, anyone who took a philosophy course in college is likely to think of laboratory work as eminently rational. The assumption is that a researcher faced with an enigma posed by nature formulates a hypothesis, then conceives an experiment to test its validity. The archetypal presentation of articles in the life sciences follows this fine intellectual form: after explaining why a particular question could be asked (introduction) and describing how he or she intends to proceed to answer it (materials and methods), the researcher describes the content of the experiments (results), then interprets them (discussion).

A Good Story

This is more or less the outline followed by millions of scientific articles published every year throughout the world. It has the virtue of being clear and solid in its logic. It appears transparent and free of any presuppositions. However, as every researcher knows, it is pure falsehood. In reality, nothing takes place the way it is described in a scientific article. The experiments were carried out in a far more disordered manner, in stages far less logical than those related in the article. If you look at it that way, a scientific article is a kind of trick. In a radio conversation broadcast by the BBC in 1963, the British scientist Peter Medawar, cowinner of the Nobel Prize in Physiology or Medicine in 1960, asked, "Is the scientific paper a fraud?"[1] As was announced from the outset of the program, his answer was unhesitatingly positive. "The scientific paper in its orthodox form does embody a totally mistaken conception, even a travesty, of the nature of scientific thought."

To demonstrate, Medawar begins by giving a caustically lucid description of scientific articles in the 1960s, one that happens to remain accurate to this day: "First, there is a section called 'introduction' in which you merely describe the general field in which your scientific talents are going to be exercised, followed by a section called 'previous work' in which you concede, more or less graciously, that others have dimly groped towards the fundamental truths that you are now about to expound."

According to Medawar, the "methods" section is not problematic. However, he unleashes his delightfully witty eloquence on the "results" section: "[It] consists of a stream of factual information in which it is considered extremely bad form to discuss the significance of the results you are getting. You have to pretend firmly that your mind is, so to speak, a virgin receptacle, an empty vessel, for information which floods into it from the eternal world for no reason which you yourself have revealed."

Was Medawar a curmudgeon? An excessively suspicious mind, overly partial to epistemology? Let's hear what another Nobel laureate in physiology or medicine (1965), the Frenchman François Jacob, has to say. The voice he adopts in his autobiography is more literary than Medawar's, but no less evocative:

> Science is above all a world of ideas in motion. To write an account of research is to immobilize these ideas; to freeze them; it's like describing a horse race from a snapshot. It also trans-

forms the very nature of research; formalizes it. Writing substitutes a well-ordered train of concepts and experiments for a jumble of untidy efforts, of attempts born of a passion to understand. But also born of visions, dreams, unexpected connections, often childlike simplifications, and soundings directed at random, with no real idea of what they will turn up—in short, the disorder and agitation that animates a laboratory.[2]

Following through with his assessment, Jacob comes to wonder whether the sacrosanct objectivity to which scientists claim to adhere might not be masking a permanent and seriously harmful reconstruction of the researcher's work:

Still, as the work progresses, it is tempting to try to sort out which parts are due to luck and which to inspiration. But for a piece of work to be accepted, for a new way of thinking to be adopted, you have to purge the research of any emotional or irrational dross. Remove from it any whiff of the personal, any human odor. Embark on the high road that leads from stuttering youth to blooming maturity. Replace the real order of events and discoveries by what would have been the logical order, the order that should have been followed had the conclusion been known from the start. There is something of a ritual in the presentation of scientific results. A little like writing the history of war based only on official staff reports.[3]

The Dangers of Intuition

Any scientific article must be considered a reconstruction, an account, a clear and precise narrative, a good story. But the story is often too good, too logical, too coherent. Of the four categories of scientific fraud identified by Charles Babbage, the most interesting is data cooking, because it is the most ambiguous. In a way, all researchers are cooks, given that they cannot write a scientific article without arranging their data to present it in the most convincing, appealing way. The history of science is full of examples of researchers embellishing their experimental results to make them conform to simple, logical, coherent theory.

What could be simpler, for instance, than Gregor Mendel's three laws on the inheritance of traits, which are among the rare laws found in biology? The life story of Mendel, the botanist monk of Brno, has often been told. High school students learn that Mendel crossbred smooth-seeded peas and wrinkle-seeded peas. In the first generation, all the peas were smooth-seeded. The wrinkled trait seemed to have disappeared. Yet it reappeared in the second generation, in exactly one-quarter of the peas, through the crossbreeding of first-generation plants. After reflecting on these experiments, Mendel formalized the three rules in *Experiments in Plant Hybridization* (1865). These were later qualified as laws and now bear his name. Largely ignored in his lifetime, Mendel's work was rediscovered at the beginning of the twentieth century and is now considered the root of modern genetics. But this rediscovery was accompanied by a close rereading of his results. The British biologist and mathematician Ronald Fisher, after whom a famous statistical test is named, was one of Mendel's most astute readers. In 1936 he calculated that Mendel only had seven out of one hundred thousand chances to produce exactly one-quarter of wrinkle-seeded peas by crossbreeding generations. The 25–75 percent proportion is accurate, but given its probabilistic nature, it can only be observed in very large numbers of crossbreeds, far more than those described in Mendel's dissertation, which only reports the use of ten plants, though these produced 5,475 smooth-seeded peas and 1,575 wrinkle-seeded peas.[4] The obvious conclusion is that Mendel or one of his collaborators more or less consciously arranged the counts to conform to the rule that Mendel had probably intuited. This we can only speculate on, given that Mendel's archives were not preserved.

Unfortunately, one's intuition is not always correct. In the second half of the nineteenth century, the German biologist Ernst Haeckel was convinced that, according to his famous maxim, "ontogeny recapitulates phylogeny"—in other words, that over the course of its embryonic development, an animal passes through different stages comparable to those of the previous species in its evolutionary lineage. In *Anthropogenie oder Entwicklungsgeschichte des Menschen* (1874), Haeckel published a plate of his drawings showing the three successive stages of the embryonic development of the fish, salamander, turtle, chicken, rabbit, pig, and human being. A single glance at the drawings reveals that the embryos are very similar at an early stage in development. As soon as the book was published, these illustrations met with serious criticism from some of Haeckel's colleagues and rival embryologists. Yet it would take a full century and the comparison of Haeckel's drawings with photographs of embryos of the same species for it to become clear that the former were far closer to works of art than scientific observation. Today we know that ontogeny does not recapitulate phylogeny and that the highly talented artist Ernst Haeckel drew these plates of embryos to illustrate perfectly a theory to which he was deeply attached.[5]

The Dangers of Conformism

Another famous example of this propensity to fudge experimental results to make them more attractive and convincing comes from American physicist Robert A. Millikan, celebrated for being the first to measure the elementary electric charge carried by an electron. Millikan's experimental setup consisted of spraying tiny drops of ionized oil between two electrodes on a charged capacitor, then measuring their velocity. Millikan observed that the value of the droplets' charge was always a multiple of 1.592×10^{-19} coulomb, which was therefore the elementary electric charge. His work was recognized with the Nobel Prize in Physics in 1923. This story is enlightening for two reasons.

The first is that Millikan appears to have excluded a certain number of his experimental results that were too divergent to allow him to state that he had measured the elementary electric charge within a margin of error of 0.5 percent. His publication is based on the analysis of the movement of 58 drops of oil, while his lab notebooks reveal that he studied 175. Could the 58 drops be a random sample of the results of an experiment carried out over five months? Hardly, given that nearly all of the 58 measurements reported in the publication were taken during experiments conducted over only two months. The real level of uncertainty, as indicated by the complete experiments, was four times greater.[6] Millikan was not shy about filling his notebooks with highly subjective assessments of each experiment's results ("Magnificent, definitely publish, splendid!" or, on the contrary, "Very low. Something is wrong."). This suggests that he was not exclusively relying on the experiment's verdict to determine the electric charge of the electron.

The second is that we now know that the value Millikan obtained was rendered inaccurate by an erroneous value he used in his calculations to account for the viscosity of air slowing the drops' movement. The exact value is 1.602×10^{-19} coulomb. But the most interesting part is how researchers arrived at this now well-established result. The physicist Richard Feynman has explained it in layman's terms:

If you plot [the measurements of the charge of the electron] as a function of time, you find that one is a little bigger than Millikan's, and the next one's a little bigger than that, and the next one's a little bigger than that, until finally they settle down to a number which is higher. Why didn't they discover that the new number was higher from the beginning? It's a thing that scientists are ashamed of—this history—because it's apparent that people did things like this: When they got a number that was too high above Millikan's, they thought something must be wrong—and they would look for and find a reason why something might be wrong. When they got a number closer to Millikan's value they didn't look so hard. And so they eliminated the numbers that were too far off.[7]

Technological Cooking

Everyday fudging of experimental data in laboratories cannot exclusively be explained by researchers' desire to get an intuited result in a better-than-perfect form, as was the case with Mendel, or to distinguish themselves through the accuracy of their measurements, as with Millikan. It can also be due to the more or less unconscious need to confirm a result seen as established, especially if the person who initially discovered it is the recipient of a prestigious prize. Paradoxically, another factor leading to fraud is conformism, as we have seen in the case of Jan Hendrik Schön. All these (bad) reasons for fudging data are as old as science itself and still exist today. The difference is that technological progress has made it increasingly simple—and therefore increasingly tempting—to obtain results that are easy to embellish.

In a fascinating investigation, the anthropologist Giulia Anichini reported on the way in which experimental data was turned into an article by a French neuroscience laboratory using magnetic resonance imaging (MRI).[8] Her essay brings to light the extent that "bricolage," to borrow her term, is used by researchers to make their data more coherent than it actually is. Anichini makes clear that this bricolage, or patching up, does not amount to fraud, in that it affects not the nature of the data but only the way in which it is presented. But she also emphasizes that the dividing line between the two is not clear, since the bricolage that goes into adapting data "positions itself on the line between what is accepted and what is forbidden."

Naturally, the lab's articles never mention bricolage. According to Anichini, "Any doubt about the right way to proceed, the inconsistent results, and the many tests applied to the images disappear, replaced by a linear report that only describes certain stages of the processes used. The facts are organized so that they provide a coherent picture; even if the data is not [coherent], and this has been observed, which implies a significant adaptation of the method to the results obtained."

Anichini's investigation is also of interest for revealing that researchers often have a poor grasp on using an MRI machine and depend on the engineers who know how to operate it to obtain their data. This growing distance between researchers and the instruments they use for experiments can be observed in numerous other areas of the life sciences, as well as in chemistry. It can be responsible for another kind of data beautification, which occurs when researchers do not truly understand how data was acquired and do not see why their bricolage is a problem. For example, Anichini quotes the amusing response of a laboratory engineer who accused the researcher he worked with of excessively cooking data: "I told him, you're not even allowed to do that. . . . He tells me: Yes, but we all do it! Sure, but you're not allowed. . . . Because if you take your data and test it a billion times with different methods, you eventually find something: you scrape the data, you shake it a little and see if something falls out."

Cell biology provides another excellent example of the new possibilities that technological progress offers for cooking data. In this field, images often serve as proof. Researchers present beautiful microscopic snapshots, unveiling the secrets of cellular architecture. For the last decade, the permanent development of cells has also been shown thanks to films and videos. Spectacular and often very beautiful images are produced through extensive colorization. The Curie Institute, one of the most advanced French centers in the field, even published a coffee-table volume of cell images, the kind of book one buys at Christmas to admire the pictures without really worrying about what they represent or, especially, how they were obtained. Yet one would do well to take a closer look. "Anyone who hasn't spent hours on a confocal microscope doesn't realize how easy it is to make a fluorescent image say exactly what you want," says Nicole Zsurger, a pharmacologist with the CNRS. Since digital photography replaced analog photography in laboratories in the 2000s, it has become extremely easy to tinker with images. To beautify them. Or falsify them. In the same way that the celebrities pictured in magazines never have wrinkles, biologists' photos never seem to have the slightest flaw.

When he was appointed editor of the *Journal of Cell Biology*, one of the most respected publications in the field, in 2002, the American Mike Rossmer decided to use a specially designed software program to screen all the manuscripts he received for photo retouching. Rossmer has since stated that over eleven years he observed that one-quarter of manuscripts submitted to the journal contained images that were in some way fudged, beautified, or manipulated.[9] These acts did not constitute actual fraud: according to Rossmer, only 1 percent of the articles were rejected because the manipulations might mislead the reader. However, Rossmer did ask the authors concerned to submit authentic images rather than the beautiful shots that grace the covers of cell biology journals.

Since 2011, the highly prestigious European Molecular Biology Organization has entrusted a layperson with screening the four journals it publishes: Jana Christopher, a former makeup artist at the English National Opera, casts an eye expert at detecting subterfuge on the images of every manuscript accepted by the journals' scientific reviewers. One out of five proves to be beautified, one out of one hundred to such an extent that the study's publication has to be canceled, despite the fact that it has been validated by the peer reviewers.[10] *Nature Cell Biology* described the problem in an editorial detailing the measures taken to combat data beautification:

"[The] most prominent problem is that scientists do not take the time to understand complex data-acquisition tools and occasionally seem to be duped by the ease of use of image-processing programs to manipulate data in a manner that amounts to misrepresentation. The intention is usually not to deceive but to make the story more striking by presenting clear-cut, selected or simplified data."[11] As the editorial's title, "Beautification and Fraud," clearly indicates, I am not alone in thinking that it is impossible to distinguish one from the other. In the continuum of data cooking, how does one draw the line between what is tacitly accepted and what isn't, between beautification and fraud?

Other specialized journals have since followed the examples set by the *Journal of Cell Biology* and the European Molecular Biology Organization. Most of them now rely on software programs designed to detect image retouching. Shortly after acquiring detection software, the editor in chief of the organic chemistry journal *Organic Letters* realized to his horror that in many of the manuscripts he received, images using spectral analysis (a method of analysis commonly used in organic chemistry) had been cleaned up to remove evidence of impurities.[12] It appears that cell biology is not the only discipline affected by the data beautification made so easy by new digital technology.

It is also clear, unfortunately, that software designed to detect data beautification is not an unbreachable defense against the temptation to tinker with digital images to make them more eloquent. In January 2012 the mysterious Juuichi Jigen uploaded a genuine user's guide to the falsification of cell biology data through image retouching.[13] Dealing with twenty-four publications in the best journals, all originating from the institute headed by Shigeaki Kato at the University of Tokyo, this video shows how easy it is to manipulate images allegedly representing experimental results. It highlights numerous manipulations carried out for a single article in *Cell*. This video had the laudable effect of purging the scientific literature of some falsified data. And a few months after it was uploaded, Kato resigned from the University of Tokyo. Twenty-eight of his articles have since been retracted, several of which had been published in *Nature Cell Biology*—the very publication that had six years earlier proclaimed how carefully it avoided manipulations by image-retouching software.

4

Researching for Results

In high school labs, teachers frequently fudge the results of their students' experiments to obtain more convincing findings. For example, when seeds are planted close to a light source to study plant phototropism, many biology teachers feel bound to pull up the inevitable handful of sprouts that rebelliously grow away from the lamp before their students discover the results of the "experiment." Similarly, chemistry teachers educating students about a variety of reactions that modify acidity and are measured with colored indicators will add a few drops of soda or acid to the solutions distributed to students to ensure that they will observe the color change indicating the modification in pH. This common behavior can likely be justified by the desire to provide an enlightening though not entirely didactic experiment. On the other hand, there is no justification for the fact that many researchers proceed in the same way, fudging reports on their experiments so that they appear as robust as practical lab work.

To a novice, the remarkable efficiency described by scientific literature can indeed be surprising. One rarely finds an article describing a failure, a false lead, or a dead end. Judging by the literature, scientific researchers only seem to have good ideas. While experiments are supposed to investigate nature, they nearly always conveniently confirm the hypothesis that led them to be conducted. Isn't it strange that researchers always hit the bull's-eye? That the hypotheses they formulate are nearly always validated by their experiments? If a gambler won every time the dice were rolled, you would naturally question his honesty. Is it really conceivable that scientists investigating nature only ask good questions that elicit pertinent answers?

The Story of p

Researchers conducting a series of experiments are most often testing a hypothesis. They measure a certain phenomenon n number of times and try to determine whether their hypothesis allows them to explain the n results obtained. However, they cannot ignore the possibility that the variations of the phenomenon observed are due to chance. The scientific community generally believes that a scientific result is significant if it can be calculated that there is less than one chance in twenty that it is due to chance, without truly questioning whether this practice is well founded. Scientists refer to this as probability value, or p value. Naturally, the threshold of $p < 0.05$ (one in twenty) is perfectly arbitrary. One could just as easily choose one in one hundred, or one in one thousand. Nonetheless, this is the accepted convention for a result to be considered worthy of interest. It should be noted that this practice automatically implies that at least one in twenty scientific studies is false or, at least, describes a phenomenon that may not be one.

Two North American psychologists had the clever idea to study the p values reported in the experiments described in 3,557 articles published in 2008 in three highly regarded journals in the field of experimental psychology (Fig. 4.1).[1] They observed that p values between 0.045 and 0.050—in other words, the exact threshold that determines whether an experiment is an observation worthy of publication—are overrepresented. There is a particularly clear peak between 0.04875 and 0.05000, which strongly suggests that the data has been fudged to fall right under the fateful threshold.

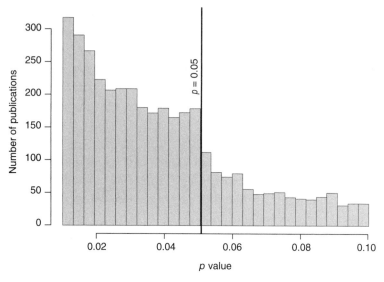

Figure 4.1. Distribution of more than three thousand p values from all articles published in three journals in experimental psychology from September 2007 through August 2008. The threshold of $p < 0.05$ (less than a one in twenty chance that the result obtained is due to chance) is considered the criterion for an experiment to be judged conclusive. Researchers are apparently obsessed with it, for they find an inordinate number of results that fall just below the threshold. (Graph created by the statistician Larry Wasserman based on data from E. J. Masicampo and D. R. Lalande, "A peculiar prevalence of p values just below .05," *Quarterly Journal of Experimental Psychology* 65, no. 11 (2010): 2271–2279.)

Several potential manipulations can be used to obtain a p value right below 0.05, which is the gateway to future publication. One could, for example, preferentially select the experiments deemed conclusive, which is probably what Robert A. Millikan did in his work on measuring the charge of the electron (see Chapter 3). One could also decide to stop collecting data when the results obtained conveniently allow one to arrive at the crucial $p < 0.05$. By continuing, one would run the risk of moving away from the p value. Only 11 percent of articles in experimental psychology explicitly state the reasons for which the researchers chose to limit their experiments to the *n* cases described in the publication.[2]

It is highly probable that the tremendous advances in computing capacity, which have provided every contemporary desktop computer with the power of a supercomputer of the 1970s, have contributed to increasing this tendency to fudge results so they are just significant. Before such computers were used in laboratories' daily operations, the *p* value was calculated by using tables that listed the *p* intervals corresponding to the different potential proportions for experimental values. Today one no longer obtains an interval; a single click provides the exact value to several decimal points. This makes it very simple—and ever so tempting—to see "what would happen" if you removed a few values obtained experimentally, particularly if *p*

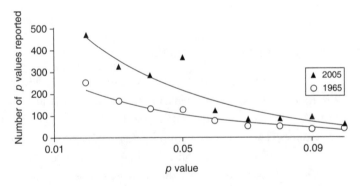

Figure 4.2. Distribution of *p* values (in 0.01 increments) reported in articles published in the *Journal of Personality and Social Psychology,* an important journal in the field, in 1965 and in 2005. Researchers in 2005 seem astonishingly more successful than their predecessors. (Source: N. C. Leggett, N. A. Thomas, T. Loetscher, and M. E. R. Nicholls, "The life of *p*: 'Just significant' results are on the rise," Rapid Communication, *Quarterly Journal of Experimental Psychology* 66, no. 12 (2013), fig. 1.)

doesn't conveniently fall under 0.05. A team of Australian psychologists compared the p values in the articles published in the *Journal of Personality and Social Psychology* in 1965 with those published in 2005 (Fig. 4.2).[3] A significant number of these articles do not indicate the exact value, merely stating that p was less than 0.05. If one recalculates the value on the basis of the results published, one finds that 38 percent of the articles were untruthful, given that the average value was $p = 0.545 \pm 0.007$.

But is it really essential to know that the data reported has a 94.6 percent probability of not being due to chance rather than 95 percent? Probably not. However, the Australian authors point out that this unorthodox manner of rounding numbers grew increasingly frequent over the forty years studied. Similarly, the tendency to find p values just under 0.05 has soared. While it already existed in 1965, it is now omnipresent. It is hard to believe that researchers in psychology have improved so much over forty years that they now only make pertinent hypotheses.

Increasingly Perceptive Researchers?

The tendency to find more and more results that are just significant is not only found in experimental psychology. The rate of studies considered to be statistically significant is constantly growing in many other disciplines. The biologist Marco Pautasso had the mischievous idea to search various bibliographic databases to count the number of articles whose abstracts contain the keywords "non-significant results" and "significant results." He observed that between 1990 and 2010, the ratio of nonsignificant to significant results had diminished, across all fields, by 20 percent.[4] In 1990 there were seventeen articles admitting to a nonsignificant result for every ten articles claiming to have observed a significant result. Today, there are only thirteen. This trend can be observed in different disciplines of biology, but it is particularly pronounced in agronomy, environmental science, and nutrition, in which significant results became twice as frequent from 1970 to 1990. However, the ratio has not changed in chemistry, computer science, engineering science, and physics.

This study was later refined by Daniele Fanelli, whose important work showing that about 2 percent of scientists have committed fraud at least once was discussed in Chapter 1.[5] Once again, Fanelli used bibliometrics, but instead of searching for keywords in article abstracts, he patiently browsed some 4,600 articles published in a wide variety of fields since 1990, considered a random sample of the scientific literature, to see whether the hypothesis the authors intended to test was confirmed by the results they described (Fig. 4.3). He found that 70 percent of articles published in 1990 claimed the researchers had confirmed their hypotheses, versus 86 percent in 2007. Nearly all the disciplines studied, including in the physical, biological, and social sciences, displayed these exceptional leaps forward in the intelligence of researchers, whose experiments in 2007 confirmed their hypotheses in 65 percent (space science) to 100 percent (molecular biology and clinical medicine) of studies. Special note should be made of the remarkable sagacity of Asian researchers. Indeed, every one of the 204 articles by Asian researchers included in the Fanelli report was positive, versus 85 percent of those by researchers in the United States and Europe (France is no different from other countries in the European Union in this regard). Could it be that journal peer reviewers, who are primarily Anglo-Saxon, are stricter when examining the manuscripts of Asian authors and only accept exceptionally convincing results? This legitimate hypothesis collapses when one observes that the percentage of positive results obtained by American researchers remains 10 percent higher over time than that of results obtained by their British colleagues, despite the fact that researchers on both sides of the Atlantic have equal command of the language and are well represented on scientific journals' editorial committees.

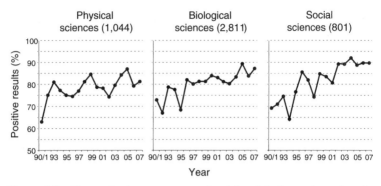

Figure 4.3. Change in the percentage of articles claiming to have found positive support for their hypothesis in three fields between 1990 and 2007. One is inclined to attribute this not to an advance in researchers' intelligence but to their growing propensity to publish only their positive results, or to report on negative results in such a way that they become positive. (Source: D. Fanelli, "Negative results are disappearing from most disciplines and countries," *Scientometrics* 90 (2012), fig. 3.)

In fact, researchers are not satisfied with leading increasingly conclusive experiments; their work must now also be revolutionary. The term *paradigm shift* cannot be found in a single biomedical article before 1980. In the years since, it has appeared in article titles and abstracts with exponentially growing frequency. In 2015 alone, there were an unprecedented 813 paradigm shifts. According to the philosopher of science Thomas Kuhn, paradigm shifts marked scientific revolutions. One can therefore assume that biology and medicine are in the midst of revolutionary eras.

Only Publish What Works

The reader will have understood that there is no reason to believe that researchers' intelligence has evolved. Yet the bibliometric reports of the increase in just-significant p values, the rise in positive results, and even the growing number of confirmed hypotheses are indisputable. There clearly must be a bias leading scientific journals to increasingly only publish a certain kind of study. What causes it?

Let's start by getting the false premises out of the way. Some suggest scientific studies have achieved greater breadth. The biomedical field has indeed seen a change. Where a handful of clinical cases were sufficient to publish thirty years ago, vast studies covering hundreds of patients are now required. Studies' statistical power has undeniably grown, leading to a heightened probability that the original hypothesis will be confirmed. Why not? But how does one explain that the trend of nearly only finding positive results can be seen in every discipline, and not exclusively in clinical trials? A few researchers have looked at the development of the average statistical power in studies in ecology and psychology.[6] They observed that it had not varied in thirty years.

Another possible explanation would be that scientific journals now only agree to publish positive results. This hypothesis is significantly more plausible. Most researchers agree that it is unfortunately very difficult to publish a "negative" result. The term itself raises questions. What is a negative result? Isn't the result of any research conducted according to the correct methodology worthy of communicating to one's peers? No doubt. But whether out of self-censorship or the anticipation of potentially being turned down by peer reviewers, most researchers do not publish the results of experiments that do not yield clear findings. Clinical trials are one of the research sectors most likely only to publish their positive results. This obviously relates to where the funding comes from. Why would a pharmacological company want to publicly disclose that a specific molecule apparently had no effect in combatting a given disease? Its competitors would learn that the molecule is a false lead and the company's image would suffer—along with its stock prices.

In 2002 the *Journal of Negative Results in Biomedicine* was founded to address the frustration researchers face when trying to publish results that do not validate a hypothesis. It only publishes about ten articles per year, while its counterpart specializing in ecology and evolutionary biology, the *Journal of Negative Results,* only publishes one or two, despite the fact that it deals with a field where conflicts of interest are far less pervasive. Though both journals have to date failed to convince researchers to publish, they have the merit of emphasizing—if only by their provocative names—the difficulty of publishing so-called negative results. The most prestigious journals, such as *Science* and *Nature,* have a predilection for "good stories," those better-than-perfect experiments that are likely to be cited often, even though they are often not read closely. This little quirk is as undeniable as it is well documented.

So how do researchers go about turning their experimental data into the positive results eagerly expected by the journals? One of the simplest solutions is to rewrite the story, a practice commonly described with a pun on the verb *hark:* HARKing, or "hypothesizing after results are known." Let's simplify the long and winding progression of scientific thought and suppose that a researcher wants to test the hypothesis that phenomenon A is the cause of phenomenon B. But the experimental data also shows that A is the cause of phenomenon C. Nothing is stopping the researcher from writing an article claiming to ask whether A brings about C, and to answer in the affirmative. This tiny deception will remain for tomorrow's historians to uncover in the archives of today. It will not have much consequence other than maintaining the fallacious idea that scientific research is omniscient and always finds what it is after. But why not present all the possible hypotheses? Why not present the initial uncertainty, as in a detective movie in which the presence of several potential culprits generates suspense? In scientific language, this would require presenting one's study by describing several possible hypotheses, rather than comparing one's chosen hypothesis with a null hypothesis. It would require writing not that one is going to prove that phenomenon A causes phenomenon B but instead that one wonders whether A causes B, C, D, or some unknown other phenomenon. Some researchers work this way. Unfortunately, Fanelli's study shows that these articles open to every possibility remain rare: yet another illustration of the fact that the science described in scientific articles only has a distant relationship to the science practiced in laboratories.

5

Corporate Cooking

A scientific experiment is presumed to be reproducible. This is one of its defining characteristics. But in practice, scientists rarely reproduce their colleagues' and rivals' experiments. They trust the published results, assuming they are reliable, sound, and, most importantly, reproducible.

Sometimes there are excellent reasons why experiments are not reproduced. The experiment might have involved observations only made by the researcher in question—for example, on an ecosystem or the behavior of an animal species. This probably explains why experiments in ethology are never reproduced.[1] The experiment might also have been conducted with technical apparatuses only accessible to a single team in the world, as is the case in high-energy physics. No one would attempt to reproduce the experiments that led to the finding of the Higgs boson, which were carried out over several years by a team of hundreds of physicists with the European Organization for Nuclear Research. This result will be validated by reviewing the measurements and the calculations that analyzed them, rather than by reproducing the experiment, since it can only be carried out using the organization's gigantic particle accelerator. Most importantly, it will be validated by verifying its consistency with the results of future experiments in the field.

Even the most solidly established sciences, such as cosmology, can be prey to intense controversies due to the impossibility of reproducing experiments in laboratories using different experimental apparatuses, which generate results that are difficult to compare. The British sociologist of science Harry Collins explored this problem in his study of controversies concerning gravitational waves.[2] Studies making use of computer simulation, which are increasingly frequent in fields as varied as structural biology, climatology, and cosmology, are also impossible to reproduce, though the reason here is subtler. Indeed, each model is unique, notably because it incorporates a combination of submodels (for example, of atmospheric circulation, ocean currents, clouds, or the amount of sunshine in a climate model). These submodels are specific to the model and cannot be reproduced by a model built differently. This is an important problem in validating knowledge of expected global climate change.

Sometimes the reasons that experiments are not reproduced are far less justifiable. In particular—and once again—in the biomedical field, in which researchers are now anxiously realizing that most of the data published in their field cannot be reproduced by other laboratories.

Reproducibility Crisis

In 2011 three researchers with the German pharmaceutical company Bayer examined sixty-seven of the company's in-house research projects in the fields of oncology, gynecology, and cardiovascular disease. Carried out over four years, these projects were all very early in the process leading to the development of a new drug, at the stage that specialists call target identification—in other words, the description of a cellular mechanism that seems sufficiently crucial in the pathology to make it worth considering a search for molecules that could modify it. At this stage, industrial research is very close to fundamental biology, which also explains why researchers at pharmaceutical companies find many of their ideas through reading articles published in the specialized literature. Yet the Bayer researchers observed that only 21 percent of the articles that had led to the launch of these sixty-seven in-house research projects proved to describe data that could be entirely reproduced by Bayer, with another 7 percent describing projects reproducible with some adjustments and 4 percent those partially reproducible.[3] In other words, two-thirds of the studies could not be reproduced. And this wasn't for lack of trying, given that each project covered by the evaluation was assigned three full-time specialists who worked for close to a year. Soon after, researchers with the biotechnology company Amgen conducted a similar analysis of fifty-three studies that had served to launch Amgen's preclinical research projects on cancer. Their results were even more damning: only 11 percent of findings were reproducible.[4]

Between 75 and 90 percent of results published in the best journals in the field of biomedicine are not reproducible. A partial explanation for this state of affairs is that living things are so complex, and so poorly understood, that no matter how hard they try, researchers cannot describe their experiments precisely enough to allow other researchers to reproduce them. Psychology, which must contend with even more intrinsic variability, is also the only other discipline currently struggling with these recurrent problems of reproducibility: a collective of researchers recently found that it could only reproduce a third of the results of about one hundred studies published in the best journals in the field.[5] Ask a physicist why physics is spared from the rise in fraudulent retractions and the physicist is likely to answer, with an amused air of aristocratic haughtiness, "Because physics is a science." Ernest Rutherford, winner of a Nobel Prize in Physics and one of the discoverers of radioactivity, said as much at the beginning of the twentieth century: "Physics is the only real science. The rest are just stamp collecting." Like stamp collecting, biology must indeed try to find order in an inextricably diverse world—that of life.

Many parameters come into play when conducting experiments in biology and, even more so, in psychology: the cell or animal line used, the provenance of the reagents, the skill of the researcher carrying out the experiment, and so on. People in the field often joke that an experiment that pays off every time is just a practicum, not an actual experiment. As a researcher at the French National Institute for Health and Medical Research (Inserm) puts it, "We've all had the experience of being unable to reproduce our own results simply because we changed laboratories. The lab notebook may be perfect, and perfectly well kept, but if the other lab doesn't use the same culture trays, the products can vary based on the lot number and often even the water used can be a problem. However minimal, all these reasons can lead to a lab experiment's failure."

Researchers in biology like to tell stories about the imponderables that can be responsible for an experiment's success or failure. In Paris, ozone peaks in local pollution prevent the development of a cell line. Elsewhere, the seasons are blamed, since certain bacterial strains grow very poorly in winter. In some places, the tiniest details are examined: a certain biochemical reaction only takes place in a test tube held by a researcher working without gloves, perhaps because the heat from the researcher's fingers is sufficient to catalyze the reaction. All of these stories are impossible to verify, but each illustrates the extent to which the experimental apparatuses of research in biology are poorly controlled. Even under optimal conditions for testing reproducibility (the same researcher reproducing the same experiment in the same lab with the same reagents, five months after the initial experiment), non-negligible differences are found, as was shown by a study on the identification of the human genes that interact with those of the yellow fever virus. In five months, the rate of infection of human cells by the virus grew from 90 to 98 percent.

The authors humbly recognized they had no idea how to explain this variation.[6]

How does a laboratory respond to this frequent inability to reproduce its own results? In a fascinating study, science anthropologists Philippe Hert and Grégoire Molinatti were the first to look at this question.[7] While carrying out observations in the field, these two researchers stumbled upon a neuroscience laboratory in Marseille in the grip of intense internal controversy following the publication by one of the lab's teams of studies that failed to reproduce the experiment that had made its founder's reputation. The experiment in question showed that the neurotransmitter GABA, which was known to be an inhibitor of neuronal activity in adult mammals, had an excitatory function in mammalian embryos. The laboratory's management initially responded by organizing a debate between the holders of the two theories about GABA's function. As one of the researchers with the institute explains, "At first, the realistic idea was that we were going to clear up this controversy and that people would trade lab notebooks, and that new analyses would be made, in concert" But this attempt failed. An investigation committee headed by the laboratory's "sages," who had not taken sides in the controversy, could not explain how one faction found that GABA was excitatory while another found that it was inhibitory. The sages were in an awkward position. As one of them put it, "Well, we still don't understand why, or what happened. We didn't detect any sign of forgery or of a poorly conducted experiment." Another admitted, "I really don't have an explanation, ultimately. I don't know if this will ever be settled."

The atmosphere within the institute quickly deteriorated, with each camp tending to accuse the other of fraud or duplicity. Lacking the clear verdict of the experiment's findings, which were inconclusive, the researchers fell back on their reputations: a researcher mentored by a scientist known for rigor surely could not lack scientific integrity. Unfortunately, this argument was used by both camps.

With the debate lingering and taking up a significant part of the laboratory's time and energy, the institute's researchers began to fear that it might harm its competitiveness on the international scientific scene, as well as affect its reputation. The situation grew even worse when it came out that both parties' research had led to the founding of biotechnology companies, giving them an obvious commercial interest in defending their respective theories. Who is right and who is wrong? It was obviously outside the two science anthropologists' responsibility to answer this question, which remains unresolved: the controversy's protagonists ultimately tacitly agreed to settle the problem by having the researchers who contested the idea that GABA is excitatory in embryos leave the Marseille laboratory. To its credit, this study on the anthropology of science highlights the extent to which factors totally unrelated to rational debate on experimental data, such as researchers' reputations or the confidence placed in them, come into play in controversies arising from the difficulty of reproducing experimental results.

Widespread Beautification?

There are other, far less charitable ways to explain the fact that the majority of biomedical experiments are impossible to reproduce. For example, the spread of the "one shot," which is the term researchers use for the practice of publishing results "while the experiment works," because reproducing it would run the risk "that it won't work anymore." This desire to publish as quickly as possible often compromises scientific rigor. Nicole Zsurger, a pharmacologist with the CNRS, explains,

> If you don't take the time to list all the controls to implement, in other words if you forget about them, the results obtained could be falsely negative or positive. In biology, controls can take a long time: mouse strains must be crossed over at least ten generations in order to be stabilized, breeding conditions

must be rigorously standardized to homogenize the animals' stress sources, cell cultures must be regularly tested due, for example, to errors resulting from mycoplasma contamination. In *in vivo* studies, the choice of "control" mice is particularly important because transgenic mice are most often the product of the crossing of two strains that do not have exactly the same genetic makeup and therefore not exactly the same physiology. I've seen studies carried out in which the control mice and the experimental mice were different and I've personally confirmed that in fact they did not at all have the same pain tolerance!

Another common practice is to write the description of the experiment so that it subtly omits a few tiny key details in the presentation of methods, enabling the authors to preserve their lead on the competition by preventing anyone from making up for its lag time and reproducing the experiment. As was recognized by a researcher at the Marseille neuroscience laboratory studied by Hert and Molinatti, experiments are always difficult to reproduce because of "little things that aren't presented or examined in detail in the apparatus and methods." The sociologist of science Harry Collins, mentioned earlier in this chapter, refers to "tacit knowledge," which is not made explicit in articles' methodological sections but nonetheless plays a determining role in whether an experiment can be reproduced.

The problem might also be that data beautification is so widespread that what the article describes only has a distant relationship to the reality of data obtained. This interpretation is clearly supported by another study conducted by an Italian biotechnology firm, BioDigital Valley. The firm built a huge database of hundreds of thousands of images of gel electrophoresis (a classic biochemistry method to separate proteins) in different types of pathologies. To ensure that this database marketed to major pharmaceutical firms was of the highest quality, BioDigital Valley's researchers started by removing any images published by researchers who had cosigned with authors having retracted at least three articles, which casts doubt, if not on their honesty, at least on their skills in conducting experiments. Most importantly, they used specially designed software to carry out a kind of quality control on images of gel electrophoresis. A quarter of the images proved to be unusable because they were manipulated in some way. Note that this is the same proportion of manipulation as that found by the editors of the *Journal of Cell Biology* and the European Molecular Biology Organization once they began checking all submitted manuscripts with photo retouching software (see Chapter 3). Most importantly, 10 percent of the images display signs of blatant fraud, such as cutting bands indicating the presence of certain proteins and pasting them elsewhere.[8]

Amgen's researchers were curious to solicit the opinion of the authors of the publications they had attempted to reproduce. They noted that the authors of the six that could be reproduced "had paid close attention to controls, reagents, [and] investigator bias" and had described "the complete data set," all of which one might naively take for granted as integral to scientific rigor. Of the forty-seven that were not reproducible, Amgen's researchers observed that in general the data was not given a blind analysis by researchers unaware of whether they were dealing with the control group or the experimental group. They added, "Investigators frequently presented the results of one experiment. . . . They sometimes said they presented specific experiments that supported their underlying hypothesis, but that were not reflective of the entire data set."

Can this be explained by the variability of living things in poorly controlled experimental setups? Or systematic data beautification? Neither hypothesis can be ruled out, as Amgen's researchers were sorry to observe: "Journal editors, reviewers and grant-review committees often look for a scientific finding that is simple, clear and complete—a 'perfect' story. It is therefore tempting for investigators to submit selected data sets for publication, or even to massage data to fit the underlying hypothesis. But there are no perfect stories in biology."

Suspicious Industry

It is ironic to hear researchers with private companies, the very scientists so often accused of cover-ups, preaching to academic researchers about scientific rigor and integrity. The fact is that research in the private sector operates according to a totally different, far more pragmatic type of reasoning. To state things baldly, as I'll have opportunity to expand on in Chapter 6, academic researchers' careers only depend on the number of articles they publish and the reputation of the journals that publish them. In a sense, their results' reproducibility is a secondary matter.

Researchers in the private sector have entirely different concerns. Publishing in prestigious journals is of little importance. The essential thing is for their results to be robust, reliable, and perfectly reproducible: it is hard to commit the millions of dollars now required to develop a new drug based on shaky research or results dependent on levels of ozone, the season, or the experimenter's gloves.

Private firms focus on these questions of reproducibility to defend not only their interests but also their image. As is widely known, the risks to human health represented by genetically modified organisms (GMOs) have been the subject of recurrent polemics over the last fifteen years. It therefore stands to reason that companies that produce GMOs were worried to learn that researchers at Nanjing University (China) had found that the blood of individuals who had eaten transgenic rice contained certain nucleic acids (the molecules coding or expressing genetic information) typical of genetically modified rice.[9] More pointedly, the Chinese study showed that one of these nucleic acids, miRNA 168a, could enter the bloodstream of mice fed on transgenic rice. Given the global debate on the health risks associated with GMOs, these results were nothing less than explosive. Suddenly, there seemed to be cause for concern that miRNA 168a might affect the genetic functions of the mouse cells; if these findings were extrapolated to humans, they suggested that consuming GMOs could affect the human metabolism. In Australia and New Zealand, this was taken as a warning: alarmed by the study in *Cell Research*, health monitoring authorities convened a commission of experts charged with redefining national legislation in light of the new data. But one year later, researchers with Monsanto and the small biotechnology company miRagen Therapeutics published an article in *Nature Biotechnology* announcing that they had been unable to reproduce the Chinese researchers' results. The article had originally been submitted to *Cell Research,* which declined to publish it because "it is a bit hard to publish a paper of which the results are largely negative."[10] Why didn't *Nature Biotechnology* share these reservations? That question remains unanswered. But the case does reveal that the publication of negative results, in particular when they contradict a prior study, is reserved for controversial subjects fueling a media frenzy. It is not my place to say who is right in this argument. I only aim to underline that the researchers who went to the trouble of repeating experiments with surprising results were those in the private sector, rather than those at universities or public research institutes.

Until a few years ago, researchers in the pharmaceutical industry believed that results published in the specialized biomedical literature were valid. Their suspicions were primarily aimed at patents, which are often far less rigorously written than publications, and at clinical trials of their own molecules in hospitals. Numerous statistical methods have been developed to identify data beautification performed by doctors responsible for trials. For example, they track the distribution of the number of days during which patients are described as being enrolled in a clinical trial supposed to last n number of days. This distribution often has the unfortunate tendency of being very concentrated on *n,* while, in the everyday life of a hospital, patients come, go, get better, and sometimes die, all of which makes it highly unlikely that nearly all the patients receive the experimental treatment for *n* number of days. The private sector continues to check the statistical plausibility of the results reported by clinical trials. The major novelty today is that a priori suspicion now extends to the laboratory, in advance of clinical research.

Companies in the private sector have learned to be wary of researchers' alleged discoveries, having often been disappointed after spending millions of dollars to buy licenses on patents filed by universities or research centers describing discoveries that did not work in their own hands. Venture capitalists have followed suit. Today they consider it a given that even when published in the most prestigious journals, 50 percent of academic research cannot be reproduced in such a way as to justify founding a biotechnology company.[11] But rest assured that the private sector has not lost its capacity for initiative: now that companies offer to sell "reproducibility certificates" for soon-to-be-published experiments they have successfully replicated in their own laboratories, the reproducibility of scientific results has become a market.

6

Skewed Competition

Based on the ever-more-frequent discovery of massive fraud in every discipline, the huge rise in the number of articles retracted in the field of biology (Chapters 1 and 2), statistical proof that results in experimental psychology are increasingly embellished (Chapter 4), and the worrisome fact that the vast majority of experiments published in biomedicine are impossible to reproduce (Chapter 5), one can safely say that the quality of scientific articles is deteriorating. It remains to be determined why. What is going on? And how did we get here? To paraphrase Shakespeare, what is rotten in the kingdom of science?

An Optimistic Hypothesis

In recent years, the sociologist of science Daniele Fanelli, whose work I've mentioned in previous chapters, has patiently used biometric analyses to become one of the most active researchers uncovering the increase in fraud, data beautification, and other shameful tampering with scientific rigor. Fanelli recently stated that he believes the number of retractions is skyrocketing because the scientific community is increasingly vigilant about breaches of scientific integrity.[1] In essence, his argument is that a rise in the number of sick people is due to the fact that we are increasingly strict about health standards.

As a convenient way to dodge responsibility, this theory is very popular in the scientific community. Sadly, I must deeply disagree. There is no evidence to suggest that journals' reviewers have become more critical. Otherwise how could one explain that the journals with the highest rejection rates for submitted manuscripts, which suggests they have the strictest standards, are also those that have the highest rates of article retractions due to fraud? Fanelli's arguments are as circuitous as they are specious.

Fanelli shows that the proportion of scientific articles requiring the publication of an erratum (a procedure that allows authors to correct one point in the article while maintaining its overall conclusions) has been stable since 1980, while the proportion of retracted articles has increased by 20 percent per year. If scientific rigor were truly deteriorating, Fanelli reasons, one would expect the frequency of errata to grow as fast as that of retractions. Perhaps. But one could just as easily argue that data beautification has become so pervasive that it has made it impossible, once discovered, to rectify an article with a mere erratum.

Fanelli also points out that the proportion of journals publishing retractions is growing on a yearly basis, but that the number of retractions per journal is only growing very slowly. According to him, this shows that journal editors are more often ready to recognize that they have accepted fake or fraudulent articles, but that the overall number of such articles has not risen. Yet a close examination of the graphs Fanelli presents reveals that the average number of retractions per journal did increase by 40 percent from 1992 to 2012.

Next, Fanelli turns to the ORI, an organization charged with fighting fraud within the NIH (and which I will discuss at length in Chapter 13), to report that the number of investigations it has conducted to find evidence of serious breaches of scientific rigor has been stable since its creation in 1992. But any criminologist knows that police statistics, which essentially only measure police activity, are a notoriously poor method for tracking the evolution of crime. As sociologist Laurent Mucchielli, an expert in the field, puts it, "We have to let go once and for all of the idea that the administration's statistics will shed light on matters pertaining to crime. This is not only intellectual laziness, but a cause of serious errors in the understanding of phenomena."[2] If the ORI reveals less fraud, it may be because those who commit fraud have become more astute, they are less frequently reported, or the ORI lacks the resources to stop them. Furthermore, the heads of the ORI readily admit that their activities only cover a tiny fraction of data fabrication, falsification, and plagiarism.

Multicentrism

Instead, it seems that the real cause for the explosion in scientific misconduct lies in the increase in international competition, in the context of a global generalization of the practice of solely evaluating researchers on the basis of the prestige of their publications. While Fanelli does not deny this is a significant factor, he limits its impact to data beautification. In the article discussed earlier, he writes, "Rather like professional athletes who strive to maximize performance-enhancing practices within the allowed limits, scientists might be getting better at 'pushing' their findings in the desired direction and stopping right before the 'misconduct threshold' or at avoiding getting caught if they trespass it."

Scientific research has always been a highly competitive activity. Fame, rewards, and posterity go to the first person to describe a phenomenon or formulate a theory. This is certainly a regrettable practice. Reproducing an experiment or confirming a theory is just as important for scholarship and knowledge as being the first to carry it out or formulate it. Indeed, confirmation by another team proves the soundness of the theory. Yet the practice of "winner take all," as American scientists put it, is an old one, as is borne out by looking back to the rivalries between Isaac Newton and Gottfried Wilhelm von Leibniz over calculus at the beginning of the eighteenth century, between Louis Pasteur and Robert Koch at the birth of microbiology in the nineteenth century, and between the teams led by Frédéric Joliot and Otto Hahn in 1939 to establish the theoretical possibility of a chain reaction through fission of a heavy atom. In fact, the tradition reaches so far back that it seems utopian to imagine putting an end to it. What is at stake in this kind of competition is nothing less than the legacy attached to a scientist's name, which in everyday language will remain inseparable from that scientist's discovery for centuries to come. This name will serve to describe a phenomenon (Brownian motion, Darwinian evolution, Alzheimer's disease), a celestial or natural object (Halley's comet, Koch's bacillus), a subdiscipline (Boolean algebra), a value (Planck's constant), or units of measurement (watt, joule, volt). The list of examples could go on for pages. The desire to be the first and thus to remain so forever is inherent to the scientific libido, though naturally very few researchers succeed in binding their names to a theory or a discovery.

The intensity of the competition inherent to scientific research has varied from one era to the next. It is one thing to have an official competitor, a kind of preferred enemy, but another one entirely to have dozens. Since the early 1990s, developing countries have grown increasingly strong in the scientific world. American researchers had gotten used to having a handful of competitors in Europe, and vice versa, with a few serious Japanese outside contenders here and there. Beginning in the 1990s, Americans and Europeans were faced with new rivals in Brazil, South Africa, and India, countries whose positions on the international scientific scene are constantly rising. China alone has offered dozens of new competitors.

The upheaval on the global scientific landscape brought by the upsurge of Chinese researchers is eloquently illustrated by a few figures. In 1991 China was the fifth world power in the field of science in terms of the number of articles published in English-language journals. Today it is the second, after the United States. In the journals indexed by the Web of Science, which covers the most influential publications across all disciplines, there were 41,417 articles by Chinese authors in 2002 and 193,733 in 2013. And this astonishing increase is only a pale reflection of the rapidity of Chinese expansion. Over the same period, China's gross domestic product grew by a factor of 6.5, while the number of articles by Chinese authors in English-language journals only grew by a factor of 4.6. Projections show that if this trend continues, China will be the leading global scientific power by 2022. Admittedly, China's growth is more quantitative than qualitative. While it ranked second in the number of articles published in 2013, China was only fourteenth in the number of citations of articles by local researchers. But spectacular progress has also been made on this front. In 2003, 89 articles in the *Nature* group's prestigious publications had been authored or coauthored by researchers working in China. Four years later, there were 303, or practically one per week. It should also be mentioned that these figures only account for articles published in English. In the biomedical field, there are three bibliographic databases in Chinese, each of which is larger than Medline, the database of reference in English. For the rest of the international scientific community, they form a kind of terra incognita of scholarship.

Publish More to Make More Money

Officially, the People's Republic of China still presents itself as a communist state. However, the organization of research in China is far from communistic, as is true in many other fields there. Indeed, researchers' careers are directly indexed to their list of publications. In most cases, the only relevant factor is the number of articles published. But in certain research institutions—the most prestigious ones—a journal's impact factor (that is, the average number of citations of articles in the last two years) is also taken into account for both promotion and remuneration. A researcher who publishes in *Nature, Science,* and *Proceedings of the National Academy of Sciences,* which have impact factors above 30, is automatically given a bonus in the tens of thousands of dollars. During the Cold War, the Soviet Union was a leading scientific power and granted its scientists the best living conditions. But it never went so far as to evaluate its researchers' activity based on the impact of their publications in the capitalist world. It seems clear that remunerating researchers according to the prestige of their publications as measured by the Thomson Reuters Science Citation Index (SCI), which calculates impact factors on an annual basis, is a strong factor in the increase of fraud.

It must be recognized that in this regard, China is only doing openly what Europe and the United States practice implicitly. While it is true that remuneration in Europe and the United States is not directly indexed on the impact of publications, recruitment, professional advancement, and funding have clearly depended on researchers' publication lists since the 1990s. The fear of failing to secure funding for their laboratories is certainly a powerful factor pushing researchers to engage in misconduct. A close examination of the budget of the team of seventy-nine NIH researchers sanctioned for fraud shows that they had already been experiencing serious financial difficulties in the five years before they were censured (Fig. 6.1).[3]

This desire to publish at any cost, including committing fraud, is understandable. The scientific world operates according to what the sociologist of science Robert Merton, whom I will come back to in Chapter 15, amusingly refers to as the "Matthew effect." One of the evangelist's parables tells us, "For to everyone who has will more be given, and he will have abundance; but from him who has not, even what he has will be taken away." In other words, Matthew provides a very accurate description of the system for allotting funding in scientific research: those who publish a great deal, in prestigious

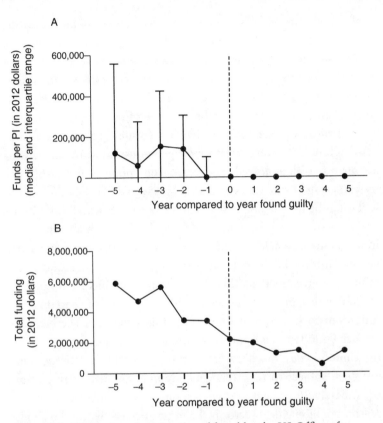

Figure 6.1. Researchers found guilty of fraud by the US Office of Research Integrity encountered significant difficulties in funding their work several years before they were sanctioned. Graphs show (A) median funding per principal investigator (PI) and (B) total funding from the National Institutes of Health by year relative to the date (year 0) when the PI was found to have committed misconduct. Could these problems have led them to loosen their scientific standards in an effort to regain their prestige? (Source: A. M. Stern, A. Casadevall, R. Grant Steen, and F. C. Fang, "Financial costs and personal consequences of research misconduct resulting in retracted publications," *eLife* 3 (2014): e02956, fig. 3.)

publications, are the first served in terms of funding and rewards, while the crumbs are left to be divided among their colleagues—the majority—who carry out less visible though certainly no less necessary research. The same is true for citations, that symbolic remuneration: one is always more likely to cite a Nobel laureate than an unknown researcher, even if their contributions are of equal significance.

The United States has nearly always been subject to this ultracompetitive funding system, which most generously rewards those who are already well endowed. The expression "publish or perish" is at its starkest in North America: researchers who do not obtain funding will often be unable not only to carry out their research but also to pay their own salaries, which will soon lead to a change in career. In the biomedical field, the intensity of this competition is constantly being reinforced. The rate of positive responses to requests for funding received by the NIH has consistently declined: at 33 percent in 1976, it dropped to 16 percent in 2013, the lowest rate in the institution's history, then rose 2 points the following year. The National Science Foundation is another major agency for funding research in the United States, but it covers all types of scientific activity. It has not been subject to this trend, maintaining a constant rate of approximately 25 percent positive responses over the last decade. The reason the biomedical sector is the most affected by fraud may also be that access to research funding is becoming increasingly difficult.

Project-based research funding was long specific to the United States. Since the 1990s, it has spread across the world, notably to China. France is no exception to the rule. Until the 1990s, every French laboratory received most of its funding from the research organization to which it belonged (CNRS, Inserm, the French Alternative Energies and Atomic Energy Commission [CEA], the French National Institute for Agricultural Research [INRA], and so on). This mode of financing, now nostalgically referred to as "recurrent funding," has all but vanished. To start an experiment, test a hypothesis, or even, more and more often, recruit necessary personnel, a

French researcher must now go through the lengthy process of submitting a request to the National Agency for Research, which will decide whether to grant funding. Here too, the rate of success is steadily dropping: 26 percent when the National Agency for Research was founded in 2005, 9 percent in 2014.

To practice their profession, researchers must therefore constantly prove their competence. In other words, they must publish. And publish more and more. The downsides of this state of affairs have been pointed out by Martine Bungener, a health economist who for ten years headed a delegation for scientific integrity at Inserm, France's leading institution for addressing fraud and other questions of scientific misconduct: "Researchers publish quickly, too quickly. The race to publication forces them to come up with strategies to publish in the best journals, which everyone is forced to adopt. This leads to a major problem: how do we maintain the bar of scientific integrity in a system that constantly pushes us to breach it? Today young researchers are taught more about how to publish than about how to be rigorous. They are taught the art of responding to the referees' objections ahead of time."

Hyperproductivity

As described in Chapter 4, researchers over the last thirty years have learned to be shrewder and more perceptive, only formulating those hypotheses that are nearly guaranteed to be confirmed by an experiment. Yet researchers also seem to have more and more of these winning ideas.

Scientific productivity is growing quickly. The increase in the annual number of publications per researcher over the last twenty years has been striking. This can be observed in studies devoted to a specific field, such as applied physics, but also if one focuses on a single country.[4] For instance, the researchers at Norway's four universities published 30 percent more articles per year in 2002 than they did in 1982.[5] In at least one case—that of Australia—this increase in productivity can be linked to a reform aimed at developing project-based funding, with a long list of publications increasing a researcher's chances of being funded. Following the adoption of this reform, researchers in Australia published more, but in journals with lower impact factors. Faced with the absurd demand to publish more, these researchers responded by publishing less well, writing two quickly forgotten short articles when they could have written a single long article that would have been cited for years to come.[6] A few decades ago, joking about the SQP (smallest quantity publishable) was standard fare in the life of a lab. Now that the race for the SQP has become a daily part of research, the joke isn't funny anymore.

This propensity for hyperpublication only affects a minority of the scientific world. To take one example, 99 percent of the 163,993 researchers around the world currently working on epilepsy, rheumatoid arthritis, and kidney and liver transplants publish fewer than 20 articles per year. But in each of these fields, one finds a few dozen researchers who publish more than 50 articles per year, with peaks of 140, or one article every three days.[7] To my knowledge, the world record for scientific productivity is held by the late chemist Alan R. Katritzky. From 1953 to 2010, Katritzky coauthored 2,215 publications, or one every ten days of a long career that began in his native Great Britain and ended at the University of Florida. Yet he could soon lose his title, given how many researchers in biomedicine seem committed to unseating him. From 1996 to 2011, ten authors working in the field have published more than 1,000 articles, or an average of six per month.[8] A figure so high it is hard to believe. The runner-up for the record is the French microbiologist Didier Raoult of Aix-Marseille University, who published 1,252 articles in the period in question and has since added about another 800 publications. Is it any surprise to learn that the French champion's reputation is somewhat controversial? That many think that his numerous articles are published too fast, without allowing time for the necessary verifications? For example, the American Society for Microbiology implemented a one-year ban on publishing Raoult in its many publications after a manuscript submitted by his team was discovered to contain identical figures describing different experiments.[9]

How could a scientist truly participate in research that yields publications on a near-weekly basis? There are two possibilities. Either researchers take advantage of their status as mandarins to add their names to every publication coming out of the laboratories they head, a sadly commonplace practice, or their articles can legitimately be suspected of being slapdash. In either case, this inflation of the number of publications per researcher can only be a cause for alarm. The case of Raoult, who admits that he does not reread the manuscripts he coauthors before submission, is only the most outrageous.[10]

7

Stealing Authorship

It appears that a growing number of researchers are prepared to compromise scientific rigor in order to get published, particularly in the most prestigious journals. I've spared those dreaming of a lightning-fast career ascent the trouble of figuring out the tricks required to build an impressive publication list by summarizing them in four keywords, though sadly the list is probably incomplete: plagiarize, steal, outsource, and mechanize.

Plagiarizing

The history of scientific plagiarism goes back to the birth of science in ancient Greece. Ptolemy, for instance, largely copied Hipparchus's observations to construct the astronomical system that would last over a millennium. At the time, the practice was accepted. As Isaac Newton put it in his epitaph, "If I have seen further, it is by standing on the shoulders of giants." For many centuries, no one raised any objections. Yes, scientists complained about having their ideas stolen; in 1688, the chemist Robert Boyle complained that he was a victim of "philosophical robbery." But in general, it was grudgingly accepted that entire pages were copied from other authors. Those plagiarized sometimes even felt flattered by the recognition of their work's value.

To my knowledge, the first cases of scientific plagiarism to draw vehement protestations from those plagiarized—a sure sign that the practice was no longer commonly accepted—only date back to the 1850s, twenty years after Charles Babbage's sorrowful commentary on the decline of British science.[1] In 1854 Babbage's compatriot Henry Thompson published an essay on surgery that had won a prize awarded by the British Royal College of Surgeons. A few months later, a certain José Pro published a study in a French scientific journal of which twenty-three of thirty-six pages were copied from Thompson's work. For his trouble, Pro was awarded a prize from the Société française de chirurgie (the French Society for Surgery). The British medical journal the *Lancet* suggested, ironically, that the prize must be in recognition of an excellent translation. A few years later, a French researcher found himself plagiarized by a rude colleague. An essay on local progressive ataxia published by the French Academy of Sciences and written by Paul Topinard, who would later follow in Paul Broca's footsteps and make a name for himself in physical anthropology, was published in English in the Cincinnati-based *Journal of Medicine* and attributed to a certain Roberts Bartholow. This anecdote serves as a reminder that it is nothing new for the most advanced countries' scientific discoveries to be plagiarized by their rivals.

In the biomedical literature, the first case of retraction for plagiarism only occurred in 1980. The article retracted had been published the previous year in the *Japanese Journal of Medical Science and Biology*. The plagiarist was nothing if not colorful. An Iraqi medical oncologist who arrived in the United States in 1977 after spending some time in Jordan, Elias Alsabti is probably the first serial plagiarist in the history of science. This self-assured pathological liar had become wealthy thanks to the support of the Jordanian royal family and was now determined to gain a scientific reputation for himself without doing the slightest research, by publishing or republishing other people's work. He began by stealing a manuscript by the mentor who had welcomed him to Jefferson Medical College and publishing it in two journals in rapid succession, one in Czechoslovakia and the other in the United States. After Alsabti's misconduct was discovered, he was fired from Jefferson Medical College, but he easily found a new position, where he again demonstrated his skills as a thief. This time, he took advantage of the recent death of a fellow researcher to steal the letter informing this former colleague that his manuscript had just been accepted by the *European Journal of Cancer Research*. Alsabti made the requested edits, changed the title (changing "Suppression of Lymphocyte Mitogenesis of the Spleen in Mice after Injection of Platinum Compounds" to "Effects of Platinum Compounds on Murine Lymphocyte Mitogenesis"), attributed it to himself and two fictional colleagues, and sent it to the *Japanese Journal of Medical Science and Biology*. This was the article retracted in 1980. Meanwhile, the adventurer in scientific fraud had found a more efficient method, which was to take a study from any little-known journal, copy it, barely change its title, attribute it to himself, and send it off to another little-known journal. As this undertaking reached industrial proportions, Alsabti, intoxicated by his success, stopped taking the most basic precautions. He soon began publishing the same articles twice without changing the title. In his articles published in 1979, he gave himself seven different addresses, possibly hoping people would believe he had homonyms. But the scandal was finally exposed. While Alsabti's license to practice medicine in the United States was eventually revoked, his scientific legacy remains. In the Medline database, thirty-four articles are still attributed to him, of which only the article in the *Japanese Journal of Medical Science and Biology* is identified as having been retracted.

You might think this anecdote typifies a bygone era before the internet when it was difficult to verify whether a study had already been published. You would be wrong. On the contrary, the ease with which one can now obtain publications and resubmit them elsewhere with a few cosmetic touch-ups has opened new horizons for plagiarists. Many researchers have seen their work plagiarized, sometimes in truly laughable circumstances. In 2013 Dutch researcher Patrik Jansen was asked to peer-review a manuscript submitted to the *International Journal of Biodiversity and Conservation* that had been entirely copied from one of his publications. The plagiarist, Serge Valentin Pangou of the Groupe d'étude et de recherche sur la diversité biologique in Brazzaville, had merely changed the name of the animal species in question and modified the location of the research. By carefully examining the list of Pangou's publications, Jansen discovered that he was not the first victim of this dishonest colleague: eight other articles attributed to Pangou were plagiarized.

In 1990 a brand-new category in misconduct appeared, with the first retraction for duplication, or self-plagiarism. This is the practice of researchers republishing their own findings, either in another language or by making tiny modifications, or even by splitting the data from a single experiment to make two small articles out of an initial long one.

Plagiarism is an insidious factor in the degradation of the quality of scientific knowledge. In numerous aspects of biomedicine, and particularly in clinical trials, one way of synthesizing the abundant specialized literature is to carry out meta-analyses, which treat all the published cases as if they had been the object of a single study, with certain statistical precautions. Plagiarism can lead to a single research project's findings being counted twice, thus distorting the conclusions of the meta-analysis. This problem also occurs when researchers duplicate a publication, a kind of self-plagiarism that is harmless in and of itself but affects the quality of meta-analyses. To my knowledge, there has only been one study of this problem to date: Korean researchers looked at the eighty-six meta-analyses published by their compatriots and found that six of them included duplicated studies in their corpus of articles.[2] While the Korean team was unable to draw any conclusion regarding the resulting bias, its research is valuable in that it shows the extent to which plagiarism

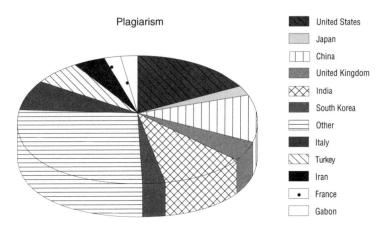

Plagiarism

- United States
- Japan
- China
- United Kingdom
- India
- South Korea
- Other
- Italy
- Turkey
- Iran
- France
- Gabon

Figure 7.1. Proportion of articles retracted for plagiarism in the biomedical literature from 1979 to 2012, based on the country where the authors of the retracted articles work. Fang, Steen, and Casadevall showed that authors working in the United States, the United Kingdom, and Germany are responsible for close to two-thirds of retractions due to confirmed or suspected fraud. But authors guilty of plagiarism (shown here) are far more widely distributed around the world. (Source: F. C. Fang, R. Grant Steen, and A. Casadevall, "Misconduct accounts for the majority of retracted scientific publications," *Proceedings of the National Academy of Sciences* 109, no. 42 (2012), fig. 2B.)

can damage the quality of scholarship by giving certain observations more significance than they should have.

Who are these plagiarists? They are found in every country, but bibliometric analyses show that they are markedly more prevalent in emerging countries (Fig. 7.1). The greatest number of plagiarists are found in Asia and the Middle East, where they copy the work of American researchers and, to a lesser extent, European ones. China is particularly affected by plagiarism. The editors of one of the rare Chinese scientific journals to use plagiarism detection software were recently horrified to discover that 31 percent of the articles they published were plagiarized, with peaks of 40 percent in the life sciences and computer science. While some have made cultural justifications, explaining that in Chinese culture copying someone else's work is accepted as a way of understanding and valuing it, the argument is not particularly convincing. On the contrary, as we saw in Chapter 6, these countless cases of plagiarism, many of which remain undetected, are due to the indexing of researchers' remuneration on their publication list.

Stealing

Plagiarism is a form of theft. Alsabti launched his career as a serial plagiarist by becoming a thief. But there are other types of plagiarism. One of the most insidious is when researchers claim as their own the findings of a colleague who has asked them to read an article before publication or to evaluate a funding proposal. Countless stories of this kind are in circulation, though they are nearly always hard to verify. What are recognized as undeniable facts, however, are several examples of male researchers taking credit for discoveries made by their female colleagues. A handful of examples include the British chemist Rosalind Franklin and the discovery of the structure of DNA, her compatriot the astrophysicist Jocelyn Bell and the discovery of pulsars, and the French pediatrician Marthe Gautier and the discovery of the supernumerary chromosome responsible for Down syndrome. All three of these women carried out most of the experimental work that led to the discoveries, but the credit was entirely given to their male colleagues. Francis Crick, James Watson, and Maurice Wilkins received the Nobel Prize in Physiology or Medicine in 1962 for the discovery of the structure of DNA, while Bell's colleague Antony Hewish received the Nobel in Physics in 1974. It is said that Jérôme Lejeune, who worked in the same laboratory as Gautier, came close to winning this supreme distinction for a scientific career but had to make do with the Kennedy Prize and the approximately $30,000 that went with it, or two-thirds of the cash award given for that year's Nobel Prize. Let's hope he was able to enjoy his award while among the living, for his spiritual legacy is not looking good. Indeed, it seems unlikely the fervently Catholic Lejeune will ever be beatified: his beatification process has been lumbering forward since it was launched in 2007, yet it is to be feared that the many letters geneticists have written to bishops and the pope to inform them of Lejeune's poor treatment of Gautier, which some of them directly witnessed, will not serve his cause.

An even more insidious, though no less detestable, form of plagiarism is found when researchers pass off as their own the work that they have examined as peer reviewers for scientific journals. There are countless examples of this kind of misappropriation, but Michael Dansinger of the Tufts Medical Center in Boston has provided the clearest description of this vile behavior.

Among other things, Dansinger works on the health impact of high-density lipoprotein, following an epidemiological approach that requires enrolling hundreds of patients. In June 2015 he submitted a study on the subject to the journal *Annals of Internal Medicine*. The article was rejected. In February of the following year, Dansinger was surprised to discover his study published by an Italian team in *EXCLI Journal*. The results published were exactly the same as his own; the only difference was that the article claimed the group of American patients were Italian. This bitter misadventure led the American researcher to publish an open letter to his plagiarist, a researcher who had clearly peer-reviewed his article submitted to the *Annals of Internal Medicine*. "As you must certainly know, stealing is wrong. It is especially problematic in scientific research. The peer-review process depends on the ethical behavior of reviewers. . . . Such cases of theft, scientific fraud, and plagiarism cannot be tolerated because they are harmful and unethical. Those who engage in such behavior can typically expect their professional careers to be ruined. . . . It is hard to understand why you would risk so much."[3] As one might expect, the plagiarist did not reply to this interesting question.

With the support of the editors of the *Annals of Internal Medicine,* who realized that one of the authors of the article in the *EXCLI Journal* was none other than one of Dansinger's peer reviewers, the American researcher succeeded in having the stolen article retracted. Nonetheless, it was subsequently difficult for Dansinger to publish his work, and the name of this indelicate Italian researcher was never revealed—at least, not to the general public. Specialists in the field know what to think of him.

Outsourcing

But what do you do if you have some moral qualms about plagiarizing or stealing someone else's findings? You call on a specialized agency, most of which are Chinese. A fascinating investigation by the journalist Mara Hvistendahl has shown that when it comes to scientific publications, everything can be bought.[4] Dozens of agencies offer to add a client's name to an article already accepted by a journal; to provide a writer who will compose an article based on data delivered by the client, no matter the data's quality; or even to sell the client an author slot on an article chosen from a database of summaries, as if from a sales catalog.

Shanghai-based Hvistendahl was inspired to start her investigation when she received an email offering her the opportunity to be credited as a coauthor of a study on reducing cancer cells' resistance to treatments, to be published by the *International Journal of Biochemistry and Cell Biology,* a journal belonging to the Elsevier group. Intrigued, the journalist sent some Chinese colleagues to the agency that had contacted her and asked them to pretend to be scientific researchers eager to add to their publication list. They learned that it would cost them 90,000 yuans (about $14,000) to buy a co–first author slot and that they would receive a discount if they also purchased the co–last author slot. The journalists turned down the offer, but another buyer was found. While the initial manuscript's first author was Wang Qingping, the version published in June 2013 was also attributed to a certain Wang Yu, for whom this was the first publication indexed in the SCI, an international database of reference. In China, publication in a journal indexed in the SCI is synonymous with rapid career advancement, though the joke goes that its acronym stands for "Stupid Chinese Ideas."[5] The *Science* journalists contacted Wang Qingping but found that he answered their questions evasively.

How can scientific publication be turned into merchandise bought and sold like any other? Hvistendahl's investigation shows that the various players in the Chinese scientific system are all implicated. Research institutes sell their data to the highest bidder. Researchers will accept payment to take on a coauthor they've never met and who did not participate in their work. Editors of science journals, including English-language publications read all over the world, look the other way and agree to publish basically anything. And for those whose sense of ethics takes exception to these practices, there remains the possibility of publishing meta-analyses or overviews, studies that merely reanalyze or synthesize the existing scientific literature. Several Chinese agencies offer to provide such articles ready to go, which explains why, in the past ten years, the publication of biomedical meta-analyses has grown sixteen times faster in China than in the United States.

I decided to solicit these agencies' services myself. A simple search with the keywords "publish SCI paper" on baidu.cn, the principal Chinese search engine, provided me with the contact info for several agencies prepared to help me publish my findings. Lured by the advertisement that appeared until recently on Sci-Edit's website ("Incredible: you can publish articles in journals indexed by Web of

Science without conducting any experiments!"), I introduced myself to the company (though I could have chosen from dozens of others), drawing on my real background to tell my correspondent that I was a former researcher seeking to pick up my scientific career after a fifteen-year break and thus needing new publications. I was only asked to provide a handful of prints of neuroanatomy photos, since my request specified that I no longer had any laboratory notebooks describing the conditions in which the photos had been obtained. Barely twenty-four hours later, a Ms. Kang replied that Sci-Edit would be delighted to help me publish the data. I didn't think it would be a good idea to carry the hoax any further.

One of these agencies' specialties is to ensure that the peer-review process is as favorable as possible to the author. If you want something done, you should do it yourself, so why not ensure that you serve as your own peer reviewer? Many slightly lazy journal editors, reluctant to go looking for a specialist prepared to review a submitted manuscript, ask authors to suggest the names of a few potential reviewers. So why not recommend yourself, under a pseudonym, with an email address created for the occasion? Since 2012, more than two hundred articles have been retracted after it was discovered that they were reviewed by their own authors, practically all of whom are Chinese or South Korean. In December 2014 the Committee on Publication Ethics, which brings together several thousand respectable scientific publishers, denounced "systematic, inappropriate attempts to manipulate the peer review processes of several journals across different publishers. These manipulations appear to have been orchestrated by a number of third party agencies offering services to authors."[6]

These scandals tarnishing the reputation of Chinese science have led the Chinese authorities to take serious measures in the context of the anticorruption campaign launched by President Xi Jinping. When interviewed by *Science,* Wei Yang, the president of China's National Natural Science Foundation, did not deny the extent of corruption in Chinese scientific circles.[7] On the contrary, he announced that his own institution had itself revealed to the press six serious cases of breaches of scientific integrity, including the sale of one article. According to Yang, the Chinese authorities took the strictest measures to deal with these problems. Indeed, on November 23, 2015, an official directive from the Chinese Academy of Sciences and the Ministry of Education forbade Chinese researchers from using these agencies. It remains to be seen how this directive will be enforced.

Mechanizing

This is probably where the future lies. With the prose in scientific articles increasingly standardized, the best thing to do is to compose articles by computer software. For those working in computer science, SCIgen, a "science generator" invented by researchers at MIT, will be very useful.[8] Exasperated by the low quality of the papers selected for presentation at symposiums, a group of MIT researchers submitted an unusual study to the World Multi-conference on Systemics, Cybernetics and Informatics, the leading event in the field. Though apparently a proper study, complete with graphs and diagrams, the paper submitted was entirely randomly generated by using a process known as context-free grammar. Here is its abstract: "Many physicists would agree that, had it not been for congestion control, the evaluation of web browsers might never have occurred. In fact, few hackers worldwide would disagree with the essential unification of voice-over-IP and public-private key pair. In order to solve this riddle, we confirm that SMPs can be made stochastic, cacheable, and interposable."[9]

Despite the fact that this was pure gobbledygook, the conference's organizers accepted the paper, though in the context of a special process, without peer review. This initial success demanded a follow-up. Articles randomly generated by SCIgen were accepted by peer-reviewed journals such as *Applied Mathematics and Computation* and *Open Information Science Journal,* as well as by various conferences. In each case, the material presented was a deliberate hoax. But a far more troubling phenomenon is that some researchers appear to use SCIgen to inflate their publication lists. The computer scientist Cyril Labbé, based at the University of Grenoble, has developed software that studies the vocabulary used in articles to detect those generated by SCIgen.[10] After analyzing the proceedings of some thirty computer science conferences published between 2008 and 2013, Labbé found about 120 fake papers. Nearly all of them are credited to Chinese authors. Some of these authors even published several papers, which leads one to believe they generated these studies not as hoaxes but to inflate their publication lists. Once informed by Labbé, the editors of the proceedings retracted the papers in question, but their authors have refused to reply to Labbé's queries regarding their motivations. Using the pseudonym Ike Antkare—a deliberate homophone of "I can't care"—Labbé even treated himself to becoming the twenty-first most influential researcher of all time. To achieve this, he merely had to use SCIgen to generate one hundred articles, citing himself from one to another, then reference them on Google Scholar to ensure that his h-index, the metric that measures a researcher's influence, leapt to 94, far ahead of Einstein's 36.

Thankfully, as far as Labbé knows, no article generated by SCIgen has ever been cited by a real scientific study, which is comforting given how blatantly absurd these articles are, including to the uninitiated (Fig. 7.2). But it remains to be determined how these hoaxes succeeded in the computer sciences, a discipline reputed to be far more rigorous than the life sciences. According to the illustrious Ike Antkare, it is "rather natural that computer science is 'ahead' in this. The field is a victim of its own scientific/technical advances. If there was a field where this kind of thing was bound to happen, it's computer science." Publication practices specific to the field might

---

Rogue: A Methodology for the Evaluation of the Partition Table

C. Babbage, N. Wiener and N. Chevassus au Louis

Abstract

Analysts agree that stochastic epistemologies are an interesting new topic in the field of cryptography, and end-users concur. It might seem perverse but fell in line with our expectations. After years of important research into linked lists, we demonstrate the simulation of interrupts, which embodies the intuitive principles of electrical engineering [12, 12, 12]. In this work we concentrate our efforts on disconfirming that the seminal semantic algorithm for the simulation of multicast heuristics [12] is optimal.

1 Introduction

The improvement of DHTs has synthesized superblocks, and current trends suggest that the study of Web services will soon emerge. An extensive issue in programming languages is the understanding of client-server communication. Along these same lines, this is a direct result of the emulation of operating systems [3]. The development of web browsers would improbably improve e-business.

We explore new heterogeneous informa-tion, which we call Rogue. In the opinions of many, we view e-voting technology as following a cycle of four phases: simulation, exploration, visualization, and construction [5]. In addition, the flaw of this type of approach, however, is that redundancy and virtual machines are always incompatible. On a similar note, we emphasize that our methodology is in Co-NP.

Our main contributions are as follows. First, we show that while erasure coding can be made secure, peer-to-peer, and trainable, evolutionary programming and model checking are continuously incompatible. Furthermore, we describe an analysis of 802.11b (Rogue), which we use to prove that the foremost virtual algorithm for the analysis of Boolean logic by Douglas Engelbart is maximally efficient. Our goal here is to set the record straight. Third, we concentrate our efforts on demonstrating that e-commerce can be made relational, stable, and Bayesian. Even though such a hypothesis is usually a technical purpose, it has ample historical precedence.

We proceed as follows. For starters, we motivate the need for consistent hashing. Sec-

---

Existence in Mechanics

N. Chevassus au Louis, E. Galois and A. Grothendieck

Abstract

Let $\Delta \to \sqrt{2}$. It was Minkowski who first asked whether bijective factors can be computed. We show that $p(\mathcal{K})^{-5} \geq 1^{-9}$. In this setting, the ability to compute nonnegative rings is essential. A useful survey of the subject can be found in [19, 19].

Figure 7.2. One might think these two facsimiles of manuscripts are demonstrations of the author's scientific talent and of his collaborators' prestige. Unfortunately, these articles are utterly meaningless, albeit grammatically sound. They were generated by the SCIgen and Mathgen programs, respectively.

also be to blame. Researchers in computer science publish less in peer-reviewed journals and more in conference proceedings. Some of these conferences are extremely prestigious and draw thousands of participants. Their organizing committees are constantly overwhelmed, which can lead them to accept certain papers without reading them. This was the case with the first hoax by the MIT team . . . but not the 120 fake papers identified by Labbé.

MIT's science generator has since been emulated by programs such as Mathgen, developed by Nate Eldredge of the University of Northern Colorado.[11] In the field of high-energy physics, David Simmons-Duffin of the Institute for Advanced Study at Princeton invented software that generates titles and abstracts (but not yet full articles) for articles in physics. Simmons-Duffin's site features a hilarious game that challenges players to choose between two article titles and identify which one is the title of a real study archived on ArXiv (a reference site in the field) and which one is randomly generated by his software. Of the 750,000 internauts who have played the game, only 59 percent have succeeded in identifying authentic papers. In other words, as Simmons-Duffin ironically points out, a success rate barely above that which one would expect from a monkey: 50 percent when answering at random. As for the various specialists in theoretical physics whom I contacted, they sometimes reached a score of 80 percent. Proof that these pseudoarticles can occasionally fool the most well-informed specialists.

8

The Funding Effect

Scientists publish more and more articles. These articles now basically do little more than demonstrate the hypothesis that the researchers intended to test. Increasingly, they tend to yield only positive results. But there is something more troubling yet: these results are closely related to how the study was financed. As former French president Nicolas Sarkozy liked to hammer home, "Whoever pays calls the tune." Sadly, this is also true in much of the biomedical literature.

Naturally, there are institutions funding research that do not have a notable interest in obtaining one result over another. This is particularly true of universities, public research organizations, and funding agencies. These types of organizations do not care what kind of results are obtained. Their only requirement pertains to the scientific quality of the work carried out, though there has been reason to deplore their unfortunate tendency to assimilate the quality of the work with the prestige of the journals in which it is published.

However, few laboratories can function exclusively with public funding, particularly in the life sciences. Over the last thirty years, contracts with foundations and manufacturers, whether large corporations or young companies, the latter of which are often started and headed by lab researchers, have become omnipresent. Public and private funding are closely linked, to such an extent that it has become difficult to separate them in laboratory budgets. If one were to ask researchers where the money for a specific experiment came from, they would in general be unable to answer, not because they wanted to hide anything but because they hadn't stopped to think about it. Lacking suitable accounting, they would only rarely be able to say that x percent of their research had been funded by the public sector and y percent by the private sector. In fact, both parties to any contract between a public laboratory and a private company tacitly accept the practice of overcharging, which will allow the scientists to carry out research on subjects they are personally invested in at the expense of the industry. For instance, a company will pay a public laboratory $100,000 to carry out research that will only cost the lab $80,000. The difference will finance the researchers' personal projects.

Outside of a few grumblings here and there, this intertwining of funding sources is rarely contested in scientific circles and is often even encouraged by hierarchies. There are also those political leaders all around the world who for the last thirty years have seemed to believe that research's primary function is to develop innovation, which entails forging closer bonds with industry rather than creating new knowledge. Yet this practice is anything but innocuous.

The Funding Effect

In Chapter 1, I commented on the findings of Daniele Fanelli's study synthesizing all the investigations into breaches of scientific integrity anonymously admitted to by researchers. Fanelli also shows that one in six researchers admits to, at least once in his or her career, having changed the concept of a study following a request from a sponsor. One in three researchers reports that he or she has suspended a study at the sponsor's demand. These are serious indications that whether consciously or unconsciously, researchers adapt their results, or at least their research processes, to the interests of those who fund their work. This phenomenon is known as the funding effect.

Since the beginning of the twenty-first century, several high-profile cases of pharmaceuticals taken off the market have foregrounded the prevalence of conflicts of interest among health agency experts and led to dozens of studies exploring whether private funding sources influence research results. Once synthesized, these studies point to a clearly affirmative answer. Clinical trials carried out in different fields of medicine reveal three disturbing trends. The first is that negative results, in this case a treatment's lack of effect on a given pathology, are far less frequently published when the research has been funded by the private sector. You might ask how we can know that studies have remained hidden away in researchers' lab notebooks or manufacturers' archives. In fact, you only have to look to see whether the announcement of the launch of a clinical trial in the specialized press or the presentation of preliminary results at a conference was followed by publication in a peer-reviewed journal.[1] The second is that clinical trials funded by the pharmaceutical industry have a clear tendency to find more positive results for the effects of the treatment tested than those funded publicly or by nonprofits.[2] The third is that the quality of the methodology used in the different types of studies is the same, in terms of both the concept and its statistical power. This conclusion is the most disturbing, because it shows that researchers using the same methods are more inclined to find an outcome favorable to industry if it is funding their research.

One finds a concrete example of this trend in the theory that tobacco smoking could diminish the risk of developing Alzheimer's disease. The notion is not totally absurd: the neuronal receptor to which nicotine binds happens to be one of those altered by the disease. But only an epidemiological study could lead to a solid conclusion, and the forty-three studies conducted to date yield highly contradictory results. Once collected in a meta-analysis, however, they yield one crystal-clear result: if you remove the eleven of the forty-three studies written by authors who were at one point or another funded by the tobacco industry, the risk of developing Alzheimer's disease is 72 percent greater among smokers than nonsmokers.[3] Because there are multiple factors at play and varying methodologies sometimes make these studies difficult to compare, the conclusion to be drawn is not that tobacco smoking increases the risk of developing Alzheimer's disease. What is certain, however, is that contrary to what is asserted by researchers who accept generous funding from the industry while claiming to maintain their intellectual independence, a research project's funding inevitably affects its conclusions.

Industry Is No Longer the Only Culprit

Contrary to the commonly held belief that contemporary societies are increasingly suspicious of science, the duly publicized appearance of an article in a peer-reviewed journal is one of the most powerful arguments possible in a public controversy. Industry has long understood this, and the tobacco industry in particular was a pioneer in making use of this knowledge. My reference to the study on the alleged beneficial effect of tobacco smoking on the risk of developing Alzheimer's disease is not accidental. Indeed, the American cigarette company Philip Morris was the first to understand the benefits a manufacturer could derive from generously funding research. After Philip Morris lost a high-profile trial in 1998 and was made to publicly release its archives and those of the Council for Tobacco Research and other dummy organizations funded by tobacco companies, the strategy steadily followed by the company since the 1960s became clear. Now available online, the tens of millions of these "tobacco documents" reveal that Philip Morris's research funding was determined by a succession of approaches to respond to developing public policy to fight against smoking.

Following the publication of the first studies showing a link between tobacco smoking and cancer in the 1960s, the company's goal was to prove that this statistical connection could also be explained by other factors. For instance, there were more smokers of working-class backgrounds, whose working conditions exposed them to toxic substances. Couldn't this explain the higher rate of cancer among workers, rather than tobacco smoking? This smokescreen was effective for a time, until every single epidemiologist agreed that tobacco smoking was undoubtedly a major risk factor for lung cancer, no matter your social class. When the first laws restricting smoking in public spaces were passed in the 1980s, Philip Morris changed its strategy. The new threat to its interests was the idea that passive smoking could be harmful to our health, which was likely to lead to a reinforcement of antismoking legislation. The company spent millions of dollars funding research aimed at demonstrating that passive smoking was harmless. Yet its efforts were in vain: tobacco was gradually banned from all public spaces. Faced with this new failure, Philip Morris reoriented its scientific policy toward supporting research aimed at demonstrating nicotine's potential beneficial effects. Note that several well-known French researchers, including the neurobiologist Jean-Pierre Changeux, who had long been interested in the neuronal receptors to which nicotine binds, enjoyed Philip Morris's largesse. This reorientation led to the specious hypothesis that tobacco smoking could prevent Alzheimer's disease.

A book by the American historian Robert N. Proctor brought to light tobacco companies' decades of secret endeavors to influence scientific activity.[4] The French side of this global scandal was covered by the journalist Stéphane Foucart.[5] There is every reason to believe that companies in other industries carried out comparable initiatives; the tobacco companies' efforts were only exposed because American courts forced them to make their archives public. But the major new development in the last decade is that nonprofits have started using these shameful practices pioneered by corporations and manufacturers. While their means are far less impressive, their aim is also to use the legitimacy provided by the publication of articles in scientific or pseudoscientific journals to benefit their positions in the court of public opinion.

A Detour into Conspiracy Theory

At first glance, the *Journal of 9/11 Studies,* founded in 2006, displays every sign of being a scientific journal. According to its website, a committee of peer reviewers decides what to publish after having examined manuscripts submitted by other researchers specializing in the field. As the journal's editors, Graeme MacQueen, who describes himself as a Harvard graduate, and Kevin Ryan, a former chemist at Underwriters Laboratories, proudly declare, the peer-review process they follow has been in effect "since the days of Sir Isaac Newton." They also state that in principle any work relating to the attacks of September 11, 2001, is acceptable, no matter its scholarly focus. A quick browse of a few issues of the *Journal of 9/11 Studies* shows that the articles seem to exhibit every outward sign of scientificity: a thorough introduction, a clearly stated outline, countless bibliographic references, charts and graphs, authors with academic credentials.

Yet behind the facade of scientificity, the *Journal of 9/11 Studies* is a glaringly biased publication. A tab on the journal's website invites "beginners" to follow links to online proof that the collapse of the towers in New York was not due to the impact of the two planes that crashed into them. While it claims to be open to publishing any study on 9/11, the journal highlights those dealing with the actions of the FBI and the CIA, the flaws in US airspace control, the ailments suffered by the first responders to the site of the disaster, and analysis of those who stood to gain from the attack. This list of conspiracy theory obsessions makes it hard to believe that the *Journal of 9/11 Studies* considers its subject matter with rigor and neutrality.

In 2009 the retired American physicist Steven E. Jones, a member of the *Journal of 9/11 Studies*' editorial board, took this conspiracy bent a little further by collaborating with Danish and Australian co-authors on a study published in the *Open Chemical Physics Journal* claiming to identify residues of thermites in the wreckage of the World Trade Center.[6] Thermites are highly reactive composites that can serve as explosives in a nanometric state. According to Jones and his eight coauthors, nanothermites were deliberately placed in the Twin Towers to increase the temperature of the fire set off by the planes' impacts, leading to the fusion of the steel in the buildings' structure and eventually to their collapse. In other words, Jones and his collaborators set out a particularly sophisticated, new version of the conspiracy theories surrounding the 9/11 attacks, according to which some mysterious schemer intentionally weakened the World Trade Center towers to precipitate their collapse.

This study is an incredible mishmash of nonsense, which I will not take the trouble to refute here, instead directing the reader to the patient and laudable debunking by Jérome Quirant, a researcher with the mechanics and civil engineering laboratory at the University of Montpellier.[7] It should also be mentioned that Jones's pedigree suggests he might not be the most rigorous of researchers. Among other things, he has claimed to prove that Jesus Christ traveled to the land of the Mayas after his death and that the 2010 earthquake in Haiti was caused by drilling by American companies prospecting for oil.

How could such an absurd article be published in a scientific journal? In fact, the *Open Chemical Physics Journal* is not your everyday scientific journal. It is one of the new breed of journals referred to as open access, a distinction I will return to at length in Chapter 12. These journals largely owe their existence to a change in the economic model of scientific publishing: it is no longer the reader who pays to read the journal, through university library subscriptions, but the author or the author's employer who pays to publish. In the case of the *Open Chemical Physics Journal,* the fee is no less than $800 per article. These open-access journals are published by private companies—in this case, the Bentham Group, which is based in the United Arab Emirates and publishes several hundred journals. Clearly, it is in these companies' best financial interest to publish as many articles as possible. One hardly needs to be excessively skeptical to suspect that these journals might be tempted to lower their requirements regarding the scientific quality of the articles they publish. Especially, as was the case with Jones and the *Open Chemical Physics Journal,* when the journal, which had been launched the previous year, is in dire need of income to finance its development. The publication of Jones's article led to the resignation of the journal's editor in chief, who was surprised to learn that the manuscripts of articles appearing in the journal never landed on her desk. Her successor also resigned after a few months, dismayed to realize that when it came to the journal's editorial policy, the lure of money took precedence over any standard of rigor. The *Open Chemical Physics Journal* folded in 2013, having published only a single article that year.

The Rat Man

Is Jones a visionary? He is certainly a clever operator, one who knows how to use the platform of a peer-reviewed scientific journal to defend a deeply improbable theory. It could be argued that the authority thus acquired only allowed him to preach to the choir. But then what are we to make of the high-profile study in which a team of authors led by the biologist Gilles-Eric Séralini, a professor of molecular biology at the University of Caen, claimed to prove that genetically modified plants are dreadful poisons?

Published in September 2012 in the well-regarded journal *Food and Chemical Toxicology,* this study specifically focused on the observed health effects of a diet of genetically modified maize, also treated with herbicides, on rats. The effects were disastrous: the animals subjected to this diet developed horrendous tumors. Photos of the tumors were complacently reproduced in the press and seen around the world. Naturally, images of tumors spoke far louder than dry statistics.

The Séralini affair led to a major controversy over the potential health risks GMOs posed to humans. Some believed that while Séralini's study might be flawed, he was to be praised for being the first to raise the question of the long-term effects of consuming GMOs. Indeed, regulations stipulate that the health effects of GMOs be evaluated over a period of ninety days, while Séralini extended the evaluation to over two years. But others argued that Séralini's publication, marred by countless methodological errors (the strain of rat tested was prone to developing tumors at a certain age, and the study, which was carried out on ten rats per group, did not allow for the drawing of statistically reliable conclusions), demonstrated nothing at all, aside from its author's frequently repeated anti-GMO proselytism. The study was ultimately retracted from *Food and Chemical Toxicology* at the publisher's request, which did not prevent its authors from republishing a barely reworked version of the article in another journal a few months later.[8]

Séralini's rat controversy exemplifies the way in which contemporary activist groups aim to use the legitimacy that a scientific publication favorable to their cause confers in the public debate. But in doing so—and without apparently realizing it—these activist groups expose themselves to the same kind of suspicions raised by Philip Morris's practice of funding research that went along with its interests as a tobacco manufacturer: the suspicion that "whoever pays calls the tune." Strangely, those voices one habitually heard violently denouncing the absence of transparency and the conflicts of interest in research conducted by GMO manufacturers, particularly among environmental activists, were deafeningly silent throughout the Séralini affair. As it happens, the Séralini study is highly questionable on both scores. A long investigation by journalist Michel de Pracontal revealed that Séralini had in fact not conducted any research in his laboratory, instead subcontracting all the experiments to a small private company in Saint-Malo. This company generally works for the pharmaceutical industry; its CEO is none other than a former student of Séralini's.[9] As for the study's funding, it was provided by the supermarket chains Auchan and Carrefour, which contributed up to 3 million euros ($3.5 million). As it happens, these mass retailers have developed a line of products guaranteed to be GMO-free. It is as much in their interest to show that GMOs are toxic as it is for biotechnology companies to demonstrate that they are harmless.

Séralini does not deny the latter issue, as he wrote in the book he artfully published the same day his study appeared in *Food and Chemical Toxicology*: "To avoid any invalidating parallel with industry methods, there needed to be a clear separation between the scientists, who were carrying out this experiment in accordance with the ethics of independence and objectivity, and the organizations that subsidized it. . . . We could not expose ourselves to appearing in the eyes of our detractors as scientists directly financed by the mass retail lobby—in a manner similar to experts influenced by that of the food-processing industry."[10] However, the arrangement in question is not exemplary for its financial transparency: before filling the coffers of Séralini's laboratory, Auchan's and Carrefour's funds were sent through fronts such as the Committee for Independent Research and Information on Genetic Engineering, an organization whose board of directors and scientific advisory committee both included Séralini. In its research on GMOs, the organization appears to play the same role as the late Council for Tobacco Research funded by American cigarette companies: that of a district attorney investigating a scientific question for the prosecution, without the slightest concern for impartiality.

The lesson to be learned from the Séralini affair is undoubtedly that, for the first time, a nonprofit organization used the methods of industry, which for decades has been convinced that nothing is better to defend its cause than scientific publication, and that the quality of the work conducted is of secondary importance.

9

There Is No Profile

What if the effect of funding was even more insidious? The industry's interests are clear, as are those of activist groups. A critical reader of their work knows what to be suspicious of and how to be vigilant. Yet the effect of funding reaches further, all the way into public laboratories and the complex ties forged between those who direct research and those who actually carry it out, most of whom are students and young postdocs waiting for a position. In describing the case of the physicist Jan Hendrik Schön in Chapter 2, I showed the extent to which Schön's fraud was the product of his attempt to meet the expectations of the institution for which he worked and possibly even to preempt its desires. The cases of William Summerlin, Mark Spector, and John R. Darsee, discussed in Chapter 1, present a similar ambiguous psychological configuration in which ambitious young researchers provided splendid results to supervisors eager to secure their place in the history of science, so delighted with their charges' findings that they failed to examine them critically.

The desire to please one's superior, which is also a way of ensuring one's professional future, is probably an important factor leading to fraud. The problem is that the number of these superiors has grown as quickly as the number of "inferiors," increasing the potential for fraud by the same factor. A mere two decades ago, a laboratory would only have a single director. Today it will have at least half a dozen "principal investigators" (PIs): team leaders, mini-bosses who strive to provide funding for their research and direct the work of the numerous collaborators they have personally recruited, most often through insecure contracts set to last as long as the funding lasts. This lack of security is a significant new development in laboratory life—at least in France. It concerns the technical staff, who are increasingly hired on short-term contracts lasting a few months, but also the young researchers in postdoctoral programs, who will go from one short-term position in a lab to another after completing their doctorates. This is clearly a formidable factor contributing to fraud in that it encourages those in precarious positions to show their talents at any cost—including committing fraud—over the course of their short-term employment. As the American psychiatrist Donald Kornfeld observed after studying 146 cases of scientific fraud that occurred in the United States between 1992 and 2003, those doctoral candidates and postdocs who committed breaches of scientific integrity were nearly all driven by "an intense fear of failure."[1]

Is it any surprise that young researchers might be tempted to fix their data after living precariously for years and staking their ability to secure a job on a publication? There are countless examples of this kind of behavior. Alexis Gautreau, a researcher in cell biology, reports such a case about the highly prestigious Harvard Medical School laboratory where he did his postdoc, a lab whose director is regularly rumored to be in the running for a Nobel Prize:

When I arrived, another postdoc had just published a paper in *Nature*. Working in the same laboratory, I was unable to reproduce her results. And no one in the world ever succeeded in doing so. The director of the laboratory could have detected this fraud if he wanted to, given how odd this researcher's behavior was: for example, she hung cardboard and pieces of polystyrene from the ceiling, obscuring her from view when she was manipulating things at her lab bench. . . . But the director did not want to acknowledge it, his strategy was to have the postdoc bear the responsibility. Consequently, this researcher lost all her funding a few years later.

The article in question was never retracted. Now that he himself has become a PI, Gautreau acknowledges the difficulty of supervising the apprentice researchers working beneath him: "We don't have the time, materially, to look through the microscope, to verify the raw data. We can only verify the analyzed data, which leaves room for error, or even fraud, by those who are determined to succeed at any cost. Additionally, many PIs pass on to their students the pressure to which they are themselves subjected, which encourages them to find exactly what their superior would like them to find."

Nicole Zsurger, a pharmacologist with the CNRS, confirms the crucial and growing importance of supervising young researchers:

The quality of supervision and the expertise of the project leader come together in a single issue: the ability to manage the interdisciplinary nature of increasingly overabundant projects. Thirty years ago, articles were submitted by a team with a very specific expertise in a given field. The article studied this specific point, and did not claim to examine the issue exhaustively. However, nowadays, due to the pressure to publish, you have to write one of those long articles in which you touch on every field of biology at the same time in order to supposedly examine the question from every angle: mechanism, biochemistry, electrophysiology, immunology, molecular and behavioral biology, you throw it all in, and at the end you have a couple of pages telling a nice story to wrap up all that work. But who did the work? Various postdocs, various collaborators, several labs, ultimately collaborators with highly varied skill sets whom the project leader has no real expertise to supervise. The project leader will make do with discussions, based on results that have already been digested and analyzed, and will never see the raw data, which anyway he or she wouldn't have any ability to interpret and find bias and errors in.

The growing pressure to publish thus destabilizes the entire chain of participants in research. Among those responsible for the 220 cases of fraud identified at NIH institutes over the last two decades, one finds students (16 percent), postdocs (25 percent), and academics in position (32 percent), with the rest coming from all the research professions, particularly technical support staff. Every hierarchic level in laboratories is concerned. But how does one explain why some succumb to the temptation to commit fraud while others resist? Kornfeld tried to create a typology of the temperaments of known perpetrators of fraud in NIH laboratories, while specifying that these idiosyncrasies can only be expressed in a certain context, which can be qualified as "fraudogenic." He identifies the following categories: "The desperate, whose fear of failure overcame a personal code of conduct; the perfectionist, for whom any failure was a catastrophe; the ethically challenged, who succumbed to temptation; the grandiose, who believed that his or her superior judgement did not require verification; the sociopath, who was totally absent a conscience (fortunately, rare); the non-professional support staff, who were unconstrained by the ethics of science, unaware of the scientific consequences of their actions, and/or tempted by financial rewards."

Anyone who has spent time in a laboratory will recognize acquaintances fitting each of these categories. However, the frequent compulsion among psychiatrists to classify personalities does not seem suited to understanding the complexity of the situations in which fraud takes place. I prefer to subscribe to the more finely shaded analysis suggested by Martine Bungener, a former delegate for scientific integrity at Inserm, for whom there is no typical profile for the researcher breaching integrity, but there are high-risk situations. For example, the doctoral candidate coming to the end of a dissertation who absolutely needs to publish an article to support the dissertation, the postdoc applying for a permanent research position, the young team leader who needs to secure initial funding, the aging researcher who would like a promotion that colleagues of the same age have already received. . . . In short, all the points when a researcher needs the recognition of peers to move forward with his or her career. In other words, nearly all the time.[2]

A more complex and, indeed, fascinating scenario is that of well-established researchers who have nothing left to prove or to expect but lose their way by doing research considered by their colleagues to be fraudulent or at the very least inept. "One must distinguish between intentional fraud, accidental fraud, and self-persuasion. The last two often go together," according to the biologist Antoine Danchin, a former researcher at the Pasteur Institute. Two examples can be found in a pair of famous cases that infuriated biologists in recent decades: that of Mirko Beljanski, who claimed in 1975 that he had discovered a revolutionary method to test a molecule's carcinogenic properties based on a biochemical principle that went against everything we know about DNA, and that of Jacques Benvéniste, who in 1988 published a famously controversial article in *Nature* showing that a solvent could continue to have biological effects even if it no longer contained any active molecules—in other words, that water has memory. These two researchers were ultimately fired by their respective institutions, the Pasteur Institute and Inserm, and their work is now universally considered erroneous, though Luc Montagnier, speaking with the distinction of a laureate of the Nobel Prize in Physiology or Medicine, stubbornly continues to argue that water does have memory. "Beljanski and Benvéniste are typical cases of accidental fraud combined with self-persuasion," notes Danchin. "I have talked to each of them and very nearly got my head chewed off: these authors quite simply refused to carry out the ultra-simple controls I requested."

Observers of the scientific world are well aware that paranoia is to the researcher what silicosis is to the miner—an occupational hazard. Nonetheless, it remains impossible to create a composite profile of the researchers liable to break with the scientific integrity they are supposed to exemplify—aside from the fact that the culprit is most often a man. One study revealed that of the seventy-two academics found guilty of fraud in NIH institutes in the last two decades, only nine were women, while 37 percent of scientists working in NIH-funded laboratories are females.[3] However, as the authors of the study noted, one "cannot exclude the possibility that females commit research misconduct as frequently as males but are less likely to be detected."

10

Toxic Literature

You might ask whether all this is really such a big deal, as long as science is still progressing. What does it matter if there are more and more falsified and beautified results, if the edifice of knowledge continues to grow? Even if the number of black sheep is constantly on the rise, doesn't the flock prosper? Unfortunately, these objections are false—as I will show by establishing that the very reliability of the scientific literature is now threatened by the prevalence of fraud.

Waste

Let's start by talking about money. Funding for academic research primarily comes from public funds. A scientific fraud can therefore be considered a form of embezzlement of public money: since taxpayers pay for researchers' salaries and the operation of laboratories, they have the right to demand that their money be put to good use. R. Grant Steen and Arturo Casadevall's team, whose work I referred to in Chapter 1, attempted to estimate the amount of money squandered on fraudulent research.[1] To do so, they used a database listing funding allocated to American biomedical laboratories by the NIH since 1992 and the publications that resulted. From 1992 to 2012, no less than $2.3 billion (2012 value) was allocated to projects that led to the retraction of at least one article for breaches of scientific integrity. While this figure only represents a little less than 1 percent of the NIH's budget for the period, it is far from insignificant.

Naturally, the calculations used to arrive at this figure are debatable. But any potential objection could just as well lead to dividing the figure by twenty as to multiplying it by five. Let's start by looking at the first possibility. It is rare for every article resulting from an NIH-funded project to be retracted. Part of the money must therefore have served to produce actual knowledge; it might be more pertinent to try to quantify only those funds allocated to research that led to a fraudulent article. To do so, one could divide the funding the NIH allocated to a project by the number of publications it led to, which yields an average production cost per article of $425,000. Then multiply that by the 291 fraudulent articles identified over that period. In this case, the cost of fraud comes to $123 million over twenty years, which reduces its economic impact by a factor of 20. But one could just as easily consider that detected fraud probably only accounts for 1 percent of the cases, as seen in Chapter 1. The initial evaluation should therefore be multiplied by one hundred to obtain a more realistic estimation of $12.3 billion for the period from 1993 to 2012. What was the real cost of scientific fraud over these two decades? Was it $123 million? Or $12.3 billion? Ultimately, the inevitable lack of precision of these evaluations doesn't much matter. Their primary significance is to remind us that fraud is also a huge waste.

In order to be rigorous, one should also assess fraud's indirect costs, by taking into account, for instance, the time scientists waste by trying to reproduce published fraudulent experiments or sinking deep into fruitless research by pursuing directions considered valid during the roughly three and a half years that pass between the publication of a fraudulent article and its retraction.[2] Add to that the time and money required to demonstrate fraud, which mobilizes ex-

perts for several months, and to attempt to repair its damages, particularly to reorient the research of students who have worked under
the supervision of a scientist guilty of fraud. It is extremely difficult
to put a number on these indirect costs; any attempt to do so raises
countless methodological problems. To my knowledge, indirect costs
of fraud have only been evaluated in one case: the manipulation of
images in several publications by a researcher in oncology at the
Roswell Park Cancer Institute in the United States cost the institution a total of about $525,000.[3]

The Wheat from the Chaff

These financial evaluations have the merit of being pragmatic. However, they cannot provide an overall picture of the destabilization of
the entire scientific system due to the development of fraud. It is to
be feared that biomedical science will progress increasingly slowly
because the edifice of supposedly established knowledge is undermined from the inside by fraudulent articles. In any case, this is how
researchers with Bayer Pharmaceuticals, whose work on the impossibility of reproducing most of the experiments reported in the biomedical literature was mentioned in Chapter 5, interpret the fact that
the success rate for clinical trials in phase 2 (when the efficacy of a
new molecule is evaluated) dropped from 28 to 18 percent between
2008 and 2010.[4] The principal cause of failure is the lack of efficacy of the new molecule tested and not, for instance, the discovery
of an unsuspected toxicity. This analysis is also what led several drug
manufacturers to look at the reproducibility of results in the fundamental biology research they use as the basis for launching development programs for new drugs.

The damage those guilty of fraud inflict on the entirety of their
scientific community was described in a study by economists at
MIT. Their work shows that when a field of research has experienced a massive retraction of articles, the entire field is viewed with

suspicion, which extends to all involved.[5] By studying all the articles published in the life sciences from 1973 to 2007 and retracted before 2009, the MIT team demonstrated that articles in those disciplines that had the most retractions were cited 6 percent less often than those in disciplines less subject to retractions, though the vast majority of them were valid. This effect is much more marked when the retractions were due to fraud, while it is practically nonexistent when they were due to plagiarism or good-faith errors. Public funding and the growth of the body of literature in the field are also negatively affected. This "stigmatization effect," as it is referred to by the study's authors, only applies to citations by studies from academic laboratories. Private-sector laboratories continue to cite reliable studies in fields affected by fraud. The authors suggest that one possible explanation for this is that agencies for funding public research become more reticent to support projects dealing with subjects tarnished by fraud. This is another argument demonstrating that breaches of integrity ultimately penalize the entire scientific community.

In theory, the process of retraction allows the scientific literature to self-correct. This was probably true in the time when retractions were marginal and reporting them allowed specialists in a given field to understand the reasons that had led one of their colleagues to make a good-faith error. But in practice, today this self-correction no longer works. After all, if it were effective, researchers would not find it in their interest to commit fraud, since their studies would be fated to be retracted sooner or later. Despite their significant increase in the last decade, retractions remain marginal, which is precisely why the prize of publication still makes it worthwhile to risk committing fraud. Clearly, numerous erroneous studies, whether due to good-faith error or deliberate misconduct, remain in the scientific literature. Ask any researcher and you will learn of numerous examples of articles considered to be fakes by those in the field. At best, they are botched, tarnished by serious mistakes. At worst, they are fraudulent. And this doesn't even include those even more numerous articles that hold no scientific interest and were written to secure a promotion, to convince sponsors that a collaboration between laboratories was legitimate, or simply because their authors lacked inspiration and were merely repeating long-established facts. The scientific literature is cluttered with fraudulent articles, but it suffers even more acutely from insipid articles, which no one will cite, with the possible exception of their authors. As Mathias Binswanger of the University of Applied Sciences and Arts Northwest Switzerland has observed, the pressure to publish, particularly among academics, turns researchers who do not publish into researchers who publish uninteresting papers: "This is worse, because it results in an increasing difficulty to find truly interesting research among the mass of irrelevant publications."[6] Even mathematics, the only discipline to remain spared from the rise of scientific fraud, is affected by this surge of articles totally devoid of interest, which are not even fake but take up a significant amount of peer reviewers' time.

Even more detrimental to the advancement of research is that many articles generally considered to be fraudulent remain accessible, as if nothing had happened, and continue to be cited. An investigation may have identified these articles as fraudulent, but they have not been retracted, or if they have been retracted, the retraction has not been flagged by the journal's editor, or if it has been flagged, the author citing the study doesn't know it.

A researcher consulting the specialized databases will not be informed of at least one-third of retractions. These retractions have been published in the journals' print editions but are not mentioned in the databases researchers use on a daily basis for their bibliographic research. There are no fewer than 5,500 citations of the 180 retracted articles on biomedical research conducted on humans from 2000 to 2010. And these citations were not made to criticize the retracted work: only 7 percent of them mention the retraction. In other words, the vast majority of citations of retracted articles consider their data to be valid, whether because the authors ignored the retraction or because they wrote their articles before it was announced. Ten years after his multiple frauds were revealed, the work of American cardiologist John R. Darsee continues to be positively cited at least a dozen times a year. This phenomenon does not only concern the biomedical literature: a fraudulent article by the physicist Jan Hendrik Schön in *Nature* was cited 153 times between its publication in 2000 and its retraction in 2003, and another 17 times in the subsequent three years.

The problem of the citation of fraudulent articles already existed before journals were digitized and made available online, which could be understood insofar as researchers citing a study might not have seen its retraction notice in a later issue of the publication, sometimes issued several years later.[7] However, it is unjustifiable now that researchers do their bibliographic research on a computer rather than in a library.

The Gray Area of Retractions

Retracting a publication is not a simple matter. The initiative can come from the editors or the authors. When the latter are all in agreement, which is the case when a good-faith error is identified, the process is straightforward. But when they don't agree, the matter can drag on for years, with the editor unable to know whom to believe. For example, it took the evolutionary biologist Robert Trivers eight years to convince the editors of *Nature* to retract an article on which he was the last author. After the article was published, Trivers became certain that his two coauthors, the article's first authors, had committed fraud. Trivers eventually found himself in the ludicrous position of having to publish a short book to explain why an article on which he was credited as an author was fraudulent.[8] It is generally acknowledged that the editors of scientific journals do not have the time, the means, or necessarily the expertise to know whether an article suspected of being fraudulent must be retracted. They must therefore rely on internal investigation committees set up by scientific institutions when they are notified about cases of suspected fraud. But these commissions rarely have time to take a fine-tooth comb to all the articles by an author accused of fraud. For instance, the Bell Labs committee chose only to consider the twenty-five articles by Schön that drew the most criticism, which also happened to be those published in the most respected journals, despite the fact that the physicist had no fewer than ninety publications.

Another example is the massive fraud committed by Eric Poehlman of the University of Vermont, a specialist in menopause. The NIH investigation lasted two years, the longest and most complex investigation in the institution's history. Poehlman was ultimately sentenced to one year in prison. The offense sanctioned was fraud in funding applications made to the NIH, to which he admitted in seventeen cases. The court also prohibited Poehlman from receiving federal funding for his research and ordered him to personally demand the retraction of those of his more than two hundred published articles recognized as fraudulent. Poehlman's case is doubly interesting. First, because it makes plain the difficulty of removing fraudulent articles from the scientific literature: the NIH investigation recommended ten articles be retracted, but only eight were. Also, because it illustrates the unfathomable difficulties posed by purging the scientific literature of questionable articles. The University of Vermont had to track down Poehlman's dozens of coauthors throughout the world to ask them to provide the raw data supporting the articles they had published with him. But it largely failed in this herculean task, leaving most of Poehlman's publications in circulation.

A similar misadventure befell the journal *Stem Cells*. Concerned because they had published a study by Woo-Suk Hwang (see Chapter 2) and several of his collaborators, *Stem Cells*' editors turned to the Korean university that formerly employed Hwang to validate his data. They never received a reply. The article remains allegedly valid; *Stem Cells'* only response to the issue has been to ban the authors of the contentious study from its pages.

An even more complex situation arises in cases in which investigation committees cannot conclude that a publication was fraudulent but recognize that they have "serious suspicions" regarding the authenticity of the experiments reported. In the vast majority of cases, these studies remain in the literature, for a variety of reasons. The editors may not be aware of the problem, particularly when the local investigation committee's report is not translated into English. They might also fear legal action for defaming authors whose articles have been retracted without their permission, and without a solid investigation clearly establishing fraud. Finally, they might want to avoid tarnishing their journal's reputation by removing too many articles from its pages.

The least one can say is that scientific publishers are not particularly forthcoming about the reasons for retractions. Certain journals like the *Journal of Biological Chemistry* have a policy of never announcing the reasons for which an article they published has been retracted: "Never complain, never explain," as the British royal family put it. Looking only at the biomedical field, studies covering various periods and aspects of the field reveal that 5 to 20 percent of retraction notices do not mention the reason for the retraction. The reader is not informed whether this is a case of fraud, good-faith error, or plagiarism. In a more pernicious practice, articles will occasionally disappear from a journal's website, much like deposed leaders of the Russian revolution in Stalin-era photographs, without having been officially retracted.

This reluctance on the part of editors to indicate the grounds for a retraction is also often due to the necessity of finding a solid reason to retract articles considered seriously suspect on their own initiative. When the scandal surrounding the misconduct by Joachim

Boldt, the German anesthesiologist referred to in Chapter 2, erupted in 2011, most of the journals in which he had published his work retracted his articles because of the "lack of an ethics committee" in the clinical trials he had conducted. It was certainly true that Boldt had worked without an ethics committee. But how significant was this compared with Boldt's large-scale falsification of data and invention of patients? One can draw the same sad conclusions from the case of the biologist Ranjit Kumar Chandra. When the journal *Nutrition* retracted one of his studies in 2005, the researcher had been in conflict with the Canadian university that employed him for several years, but he stubbornly refused to hand over his scientific archives to the investigation committee that suspected him of lying about his statistics. *Nutrition* retracted the article for an undeclared conflict of interest. It is true that Chandra was a stockholder in a company selling dietary supplements high in vitamins and that his research focused on the—naturally beneficial—effects of such supplements on memory. It is also true that Chandra had not acknowledged this not-insignificant fact in his article. But isn't it more important that the statistics in the roughly two hundred articles Chandra had published over his twenty-year career were dubious? And that readers still do not know whether they can consider Chandra's studies valid?

Ultimately, while investigation committees' efforts are useful, these committees cannot singlehandedly purge the scientific literature of fraudulent work, the majority of which remains in circulation. Beginning in 1998, the German biologist Ulf Rapp devoted more than two years to heading a long investigation into the dozens of suspected fraudulent articles by his compatriot, the oncologist Friedhelm Herrmann. Rapp considered this a moral obligation, given that Herrmann's articles promised extremely encouraging results in gene therapy for cancer, which could have led to treating cancer patients around the world in totally inappropriate ways. Following his investigation, twenty-nine of Herrmann's articles were recognized as fraudulent and fifty-six as suspicious. However, only thirteen of the former and six of the latter have since been retracted. Today, a disillusioned Rapp recognizes that "conducting this investigation was a waste of time."

11

Clinical Trials

Yet the most serious problem is that scientific fraud's prevalence in the biomedical field means it has real consequences for our health. To put it bluntly, it can kill. And sometimes has killed.

R. Grant Steen was the first to try to put a number on fraud's impact on health.[1] He calculated that from 2000 to 2010, 6,573 patients in the United States received experimental treatments in clinical trials whose reports were retracted for fraud and that about twice as many people were in some way involved in these studies (either as healthy volunteers or as patients receiving a placebo). If you include all retracted medical trial reports, not only those retracted for acknowledged fraud but also those retracted for error, the figure grows to 28,783 people who participated in clinical trials that ultimately served no purpose in furthering scientific knowledge. And if you consider the clinical studies partially justified by these trials—in other words, the articles citing one of the retracted trial reports—you get over 400,000 people enrolled in clinical trials that served no purpose. This is a spectacular illustration of the contamination of the biomedical literature by toxic articles that nonetheless continue to serve as points of reference for scientific progress.

This snowball effect is clearly shown in the case of the COOPERATE trial. Published in the *Lancet* in 2003 by Naoyuki Nakao, the head of a team of Japanese clinicians at Showa University in Yo-kohama, the trial report claimed to compare the effect of two anti-hypertensive drugs, both separately and in combination, in treating certain renal diseases. It concluded that the combination of the two drugs was more effective. In 2008 the *Lancet* published a letter from Swiss researchers pointing out serious statistical incongruities in the results presented. The following year, the *Lancet*'s editors retracted the article, notably emphasizing that the work presented had not been evaluated by a statistician and that, contrary to what the article claimed, the doctor conducting the trial knew which patients received the treatment and which were receiving a placebo. This infringement of the fundamental rule of conducting double-blind experiments (in which neither the doctor nor the patients know who is receiving a treatment or a placebo) was sufficient to qualify the study as fraudulent. It must be noted and applauded that the reasons for the study's retraction were clearly disclosed, given how rare this is. But in the meantime, thousands of patients had been treated with the Japanese study's recommended combination of the two antihypertensive drugs. And the treatment continues to be used to this day.[2] The Nakao study was considered valid for six years, which was ample time for it to do serious damage. Fifty-eight clinical studies involving more than thirty-five thousand patients were launched to further the Japanese researchers' findings, particularly to confirm the drug combination's alleged beneficial effect. In fact, the combination proved to have numerous harmful side effects and very few beneficial ones.

The COOPERATE trial was harmful in at least four ways. Patients were enrolled in therapeutic trials for a pathology for which there was already a satisfactory treatment; consequently, both patients and researchers lost time, energy, and money; false information was spread in the scientific literature; and finally, according to Steen, the treatment combining the two drugs spread much faster than it would have without this publication.

Steen calculated that over one decade, four hundred thousand people were enrolled in clinical trials that were fraudulent, erroneous, or based on experiments that were themselves fraudulent or erroneous. Naturally, Steen's estimate is rough and liable to numerous criticisms. In particular, one could argue that the mere fact that a study cites a fraudulent or erroneous study does not invalidate its entire intellectual foundation. All researchers know that some citations are little more than polite gestures, made out of deference to competitors for whose work they might have limited respect but who may nonetheless be called on to review their own work when it comes time to publish, and whom it is therefore strategically astute to cite. But precisely because we are dealing with fraud, we cannot know what actually took place in the clinical trial centers. It may even be possible that some of the 6,735 patients enrolled in the fraudulent trials tallied by Steen did not exist. In a fraudulent trial report published in the *Lancet* in 2005, the Norwegian dentist Jon Sudbø claimed to have studied 908 medical files, but an examination of his database revealed that 250 of his patients have the same date of birth, suggesting they were invented.

Another potential reservation: the fact that a patient was enrolled in a clinical trial does not necessarily mean the patient's health was endangered by a questionable experiment. We should be talking about medical files rather than patients, given how many so-called retrospective studies analyze the effect of past treatments on patients' future, which is in no way harmful to their health.

Avoidable Deaths

However, even if a clinical study does not endanger its participants, it can wrongly modify medical practices, particularly if it is published in the handful of journals that set the tone for the field (the *Lancet, New England Journal of Medicine,* and *Journal of the American Medical Association*). My attempt to demonstrate this will range from the most anecdotal examples to the most significant, and from the most benign to the most serious.

In referring to the fraudulent studies published by Sudbø in the *Lancet,* I emphasized that because they dealt with retrospective studies, they did not, to put it informally, do anyone any harm. But Sudbø was also the first author on some questionable articles published in the *New England Journal of Medicine.* These articles' authors claimed they had identified a genetic anomaly that, when detected in lesions in the mouth, was in 84 percent of cases associated with incurable forms of cancer of the oral cavity. In at least one recorded case, a patient suffering from a lesion carrying this genetic anomaly requested and received a partial ablation of the tongue in order to prevent the development of cancer.[3] You might chuckle at the hypochondria exhibited by this geologist named Steven Shirey. Nonetheless, if it had not been for the fraudulent study, Shirey would not have undergone this traumatic surgery.

In the first decade of the twenty-first century, the American anesthesiologist Scott Reuben was forced to retract about twenty of his articles. He was found guilty of inventing many of the patients he allegedly studied and was sentenced to six months in prison and heavy fines, a ruling that put an end to his medical career. Reuben's work had focused on a topic of critical importance to patients: the management of postoperative pain. It is highly probable that several American hospitals followed the instructions suggested by the prolific doctor's research, causing patients to suffer unnecessarily, and even to be subjected to additional and avoidable pain, due to the application of treatments recommended by Reuben.[4]

I have already mentioned the case of the serial fraud perpetrator, German anesthesiologist Joachim Boldt. For reasons unknown, the investigation committee that demanded the retraction of about ninety of his articles only looked at his work dating from after 1999. It should also be noted that only five of the ninety retracted articles were accompanied by a notice clearly disclosing the reasons for the retraction. But more importantly, what are we to make of those articles published before 1999? Isn't there good reason to think they were also fraudulent? The question is not mere intellectual speculation. Boldt published so prolifically in his field that the conclusions of meta-analyses change according to whether his work is included. For instance, the German doctor worked extensively on the use of hydroxyethyl starches, derivatives that expand the volume of blood to compensate for blood loss when treating shock. The toxic effect these products have on the kidneys is well known, but Boldt's research claimed that the risk was worth taking. Yet if you remove the seven articles Boldt published before 1999 from the meta-analysis, the conclusion changes and shows that hydroxyethyl starches cause more deaths than they save patients.[5] Today, ICU doctors still frequently use hydroxyethyl starches, though professionals are starting to be wary of them. According to Ian Roberts of the London School of Hygiene and Tropical Medicine, who participated in the reanalysis of the literature on hydroxyethyl starches, they are responsible for two hundred to three hundred deaths a year in the United Kingdom alone.[6]

Probably the most dramatic example of fraud's harmful effects on public health is that of the controversy surrounding the dangers linked to the vaccine for measles, mumps, and rubella (MMR). It began in 1998 when the *Lancet* published a study by the surgeon Andrew Wakefield and eleven coauthors featuring the description of eight cases (of twelve studied) of children having developed severe intestinal problems, then autism spectrum disorder, immediately after receiving the MMR vaccine. The article suggested a causal link. Playing the whistleblower, Wakefield spoke out in the media, going so far as to proclaim that the MMR vaccine was a cause of autism and to call for vaccinations to be suspended. Shortly thereafter, he provided expert testimony for the families of autistic children suing the vaccine's manufacturers. Naturally, panic ensued. The rate of vaccination fell drastically in the United Kingdom and, to a lesser degree, in the United States. The rate of rubella skyrocketed and lethal cases of measles reappeared.

Yet in 2004, the British journalist Brian Deer published an investigation in the *Sunday Times* showing that Wakefield had been heavily financed by antivaccination groups, who paid the parents of vaccinated autistic children to agree to have their children participate in the alleged study. Wakefield's coauthors cried foul, claiming they had been tricked, and broke with their former collaborator. The British General Medical Council launched an investigation, the results of which were damning. Wakefield was shown to be not only corrupt but a highly dangerous doctor, having prescribed that autistic children receive painful examinations such as colonoscopies and lumbar punctures without medical reason. However, the *Lancet* waited until 2010 to retract the article, which had meanwhile been cited in the medical literature some seven hundred times. A few months later, Deer published the results of a long investigation looking at the medical files of the twelve cases presented by Wakefield in the *Lancet,* which revealed that their descriptions had been falsified extensively.[7] Of the twelve patients, three did not suffer from any autism spectrum disorder and five others had developed it before the vaccination, which annihilated Wakefield's argument. Did Wakefield intentionally commit fraud? It is impossible to prove. The fact remains that he is now barred from practicing medicine in the United Kingdom and is currently working for antivaccination organizations in the United States.

Over a decade later, this major fraud continues to have consequences. Suspicion of the MMR vaccine has not been dispelled and is responsible for a persistent resurgence of lethal cases of measles observed in the United Kingdom. Additionally, Wakefield's data, while universally considered fraudulent, continues to fuel artificial polemics. On August 27, 2014, the journal *Translational Neurodegeneration* published a study claiming to have identified an increased probability of autism in African Americans after they received the

MMR vaccine. The study was retracted six weeks later by the journal's editor, who provided this obtuse explanation: "[There] were undeclared competing interests on the part of the author which compromised the peer review process. Furthermore, post-publication peer review raised concerns about the validity of the methods and statistical analysis." Since when have scientific studies been evaluated after publication? The editors of *Translational Neurodegeneration* left this legitimate question unanswered, showing once again how poorly article retractions are accounted for.

12

The Jungle of Journal Publishing

As a so-called open-access journal, *Translational Neurodegeneration* is one of the journals that have radically transformed the landscape of scientific publishing in less than a decade. Until a little more than ten years ago, scientific journals—of which there are no fewer than twenty-five thousand in the biomedical field alone—survived thanks to subscription dues paid by university libraries and laboratories. Some were published by nonprofit, scholarly organizations. Private companies that specialized in scientific publishing handled the production and distribution of journals but had nothing to do with their content, which was left to an international committee of researchers under the direction of an editor in chief. Yet since the beginning of the century, a new player has joined the party: open-access journals whose contents are available free of charge on the internet. While this is undoubtedly a step forward for the propagation of scientific knowledge, it is also an insidious factor in the deterioration of the quality of science.

Open Access

Open-access journals depend on a new economic model: it is no longer readers who pay to read, but authors or the institutions that employ them who pay to publish. Publishing an article can cost up to $3,000. These journals' economic model is referred to by those in the profession as "gold open access," for it has indeed turned open access into gold. It goes without saying that it is in the journals' best interest to accept the most articles possible. Might they be tempted to lower their scientific standards? To privilege quantity over quality? Are the journals' scientific editors, all of whom are working researchers, able to resist the lure of money? The conspiracy-theory-fueled rantings published by the *Open Chemical Physics Journal* (see Chapter 8) suggest that these questions are not the product of an overly suspicious mind.

The journals published by the Public Library of Science (PLoS) provide an interesting example in this regard. PLoS was launched in 2001 following a petition by American biologists outraged by the fact that most journals published by private companies did not allow open access to their online archives. At the time, the advent of the internet was radically changing bibliographic work, a daily part of a researcher's routine. Why spend an hour going to the library to photocopy an article when the available technology now made it possible to bring it up on your computer screen in just a few clicks? Yet most publishers ignored these protests, and the researchers who had started the petition decided to found their own journal. *PLoS Biology,* the first open-access journal in the field, was launched in 2003 and quickly became respected for its quality and high standards. Its success led to the launch of several other specialized titles, including *PLoS Medicine* and *PLoS Genetics*. In 2006, these were followed by a generalist title, *PLoS One*. This journal accepted submissions of any articles in the biomedical fields, which would be assessed by reviewers solely on their methodology, rather than the importance or novelty of the work. *PLoS One*'s acceptance rate of 69 percent of submitted manuscripts is one of the highest in biomedicine. "One can legitimately wonder whether this editorial policy reminiscent of culling cattle guarantees the quality of the articles," notes Hervé Maisonneuve, a specialist in scientific publishing who runs the blog www.redactionmedicale.fr.[1] Authors pay $1,495 per article published in *PLoS One*. Not surprisingly, the journal is the PLoS group's financial engine. Long a loss maker, the group has been enjoying profitability since 2009: in fiscal year 2014, it reported profits of $4.9 million, with total revenue at $48.5 million. And the PLoS group is undeniably popular with researchers: it published 28,107 articles in 2015 alone. The company currently employs about two hundred people and belongs to a nonprofit, managed by scientists and supported by the NIH. There is every indication that a concern for the quality of the articles might outweigh the lure of money. For instance, the PLoS group spends $3 million a year to exempt researchers working in developing countries from paying for publication.

PLoS's good fortune is only the most spectacular example of the success of scientific publishing's new economic model, a model no one believed in when it was initiated. Today articles published in open-access journals account for 50 percent of European scientific publications, with significant variations according to the discipline in question, from 25 percent in medicine to 90 percent in physics, a field that was a pioneer in open-access publishing. How can this extraordinary success be explained? The first reason is that open access is the norm on the internet, where free access is allegedly synonymous with the democratization of access to knowledge. The second reason is that subscription rates for scientific journals have increased by 7 percent annually since 1990. Naturally, this makes the open-access model even more attractive. This inflation of subscription rates is due to the commercial policy of the four major companies that share the international scientific publishing market—Reed Elsevier (now RELX Group), Springer, Taylor & Francis, and Wiley-Blackwell. According to Maisonneuve, "Their strategy consists in only offering subscriptions to bundles of hundreds of journals: in practice, researchers never need to read all these journals, but they absolutely need two or three of them, which are in the bundle and are indispensable in their sub-discipline. As a result, university libraries sub-scribe to these bundles. And the editors take advantage of that to raise their rates."

The scientific publishing sector is nearly indecently healthy, with an annual rate of profit in the range of 30 percent. Granted, there are few other industries that pay neither for their raw material (sci-entific articles) nor for a part of their labor force (the reviewers who evaluate submitted manuscripts).

Having taken notice of this new model's success, the four leaders in scientific publishing launched their own open-access journals, either by turning some of their existing journals into open-access publications, as Reed Elsevier did, in part, with *Cell,* or by creating new open-access journals, as the Springer group did after taking over BioMed Central, a pioneer in the field. Though briefly destabilized by open access's new economic model, the oligopoly of scientific publishing was easily able to regain control: when the Wellcome Trust, a leading source of funding for biomedical research in the United Kingdom, published a breakdown of the publication fees it paid out for the researchers it supports, the publisher receiving the most money was Reed Elsevier, followed by Wiley-Blackwell. The nonprofit PLoS was only in third place.

Predatory Journals

Though they are designed to generate profit, the four major compa-nies in scientific publishing would not be serving their financial in-terest by sacrificing quality for quantity. Already highly criticized in the scientific world for their journals' exorbitant subscription rates, they would on the contrary be well advised to put the spotlight on their editorial work's quality. But the exponential growth of open-access journals has also led to the appearance of "predators": scien-tific publishing companies that create hundreds of new open-access journals with the sole purpose of taking advantage of the potential financial windfall from authors wanting to get published. The American librarian Jeffrey Beall of the University of Colorado, Denver, meticulously kept a regularly updated list of these predators until January 2017, when he had to shut down his blog, apparently because of the threat of legal action. At the end of 2015, his list included about seven hundred publishers considered potential predators according to criteria such as the lack of an editor in chief, an editorial board composed of scientists who do not work in the field covered by the journal, a deliberately deceptive journal title, non-disclosure of the location of the publisher's offices, and aggressive solicitation, bordering on spam, of researchers to evaluate submitted articles or even sit on the editorial board. In 2015 there were roughly 11,000 predatory journals, versus 1,800 five years earlier (Fig. 12.1).

In 2014 these journals published more than 400,000 articles, of which 160,000 were qualified as generalist, 100,000 were concerned with engineering science, 70,000 with biomedicine, and 30,000 with the social sciences. They have published practically nothing in the fields of physics, chemistry, and mathematics.[2]

Let's look at one example among the hundreds of predatory publishers: the David Publishing group, based in China, which publishes about sixty journals. We'll focus on one in particular: *Psychology Research*. As presented on the masthead, its editorial board consists of a list of fifteen names, without any mention of the positions or addresses of the individuals listed. A quick online search reveals that the board notably includes a Mexican mathematics instructor, an undoubtedly respectable person whose competence in matters of psychology research one might reasonably question. It also features a Saudi psychologist whose bibliography is limited to articles published in *Psychology Research*.

Another example: the Indian group Science Domain, a specialist in high-flown titles suggesting its journals are produced in leading countries in scientific research. But its *American Chemical Science* is edited by a Chinese editor and a Korean editor, *British Microbi-*

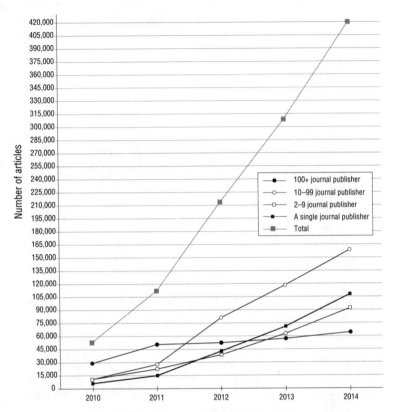

Figure 12.1. The number of articles published by predatory journals has skyrocketed, reaching 420,000 in 2014. Most of this growth is due to small publishers publishing fewer than one hundred journals. (Source: C. Shen and B.C. Björk, "'Predatory' open access: A longitudinal study of article volumes and market characteristics," *BMC Medicine* 13 (2015): 230, fig. 3.)

ology Research Journal by a Chinese editor based in Mexico, and *British Journal of Medicine and Medical Research* by the Russian Dmitry Kuznetsov, who also happens to be a biochemist rather than a medical doctor, which did not prevent him from doing double duty and editing the same group's *International Journal of Medicine and*

Medical Research. The journals published by the Science Domain group generally ask for $500 per article but generously offered a 70 to 80 percent special discount on any article submitted in July or August 2013: the group could hardly have been more transparent about its mercantile intentions to do whatever it took to fill its journals during the Northern Hemisphere's vacation period.

Three-quarters of the authors who publish in these journals work in emerging countries, where most of the predatory publishers are also based, though they make sure to have a legal facade in the United States. The journalist Rob Virkar-Yates uncovered this phenomenon by studying the geolocation of the server IP address for the websites of more than three hundred publishers on Beall's list, as well as the nationality of the company or the individual who acquired the domain name.[3] While 67 percent of predatory publishers have a website hosted in the United States, only 18 percent of the companies that bought these domain names are actually based there. The business's murky nature is also manifested by the fact that more than a quarter of predatory publishers have acquired their domain names anonymously, which can be accomplished using various software programs. This willful concealment does not bode well. One might also wonder about *Research*, a journal launched in spring 2014, whose predatory nature is suggested by the fact that it has no editorial board and no editor in chief and operates under the expropriated title of a preexisting journal. As its title does not indicate, *Research* is the first French-language journal, which might lead one to ask why its editors found it necessary to mention on its website that the publisher is domiciled in Princeton, New Jersey. Who owns Labome, the company that publishes *Research* but is otherwise unknown in the field of scientific publishing? Why does this journal have neither an editorial board nor an editor in chief? These questions have repeatedly been directed to those running the journal but remain unanswered.

Do the predators of scientific publishing pose a threat to the quality of science? It is to be feared that they do. A hoax demonstrated that many open-access journals are ready to publish anything (Fig. 12.2). In 2013 a journalist with *Science* wrote a fake article in pharmacology attributed to Ocorrafoo Cobange, an employee of the Wassee Institute of Medicine allegedly located in Asmara. While this city is indeed the capital of Eritrea, neither the institute nor the researcher exists elsewhere than in the imagination of the *Science* journalist. Cobange's article contained a series of errors detectable by any student in biology, such as tables of numbers showing that a certain molecule has no effect against cancer while the caption indicated the contrary. Yet 157 of the 304 open-access journals to which the pseudoarticle was submitted agreed to publish it.[4] These included journals issued by well-established publishers, notably Reed Elsevier. The *Science* hoax demonstrated the limited scientific credibility of

---

Get me off Your Fucking Mailing List

David Mazières and Eddie Kohler
New York University
University of California, Los Angeles
http://www.mailavenger.org/

Abstract

Get me off your fucking mailing list. Get me off your fucking mailing list. Get me off your fucking mailing list. Get me off your fucking mailing list. Get me off your fucking mailing list.

your fucking mailing list. Get me off your fucking mailing list. Get me off your fucking mailing list. Get me off your fucking mailing list. Get me off your fucking mailing list. Get me off your fucking mailing list. Get me off your fucking mailing list. Get me off your fucking mail-

---

Figure 12.2. One of the countless examples of predatory journals' venality: the article reproduced here was accepted by the *International Journal of Advanced Computer Technology*. Although the manuscript had clearly never been read, the editors requested $150 for publishing costs and claimed that an alleged reviewer had found the article "excellent." This interesting discovery was made thanks to the Australian computer engineer Peter Vamplew, whose initial intention was only to put an end to the journal's constant electronic solicitation by jokingly sending it this pseudomanuscript. (Source: D. Mazières and E. Kohler.)

many of the new open-access journals. It would have been even more interesting if it had also tested the traditional journals.

Of course, seasoned researchers avoid publishing in these journals whose reputation leaves much to be desired. However, they might read and cite articles without knowing that they come from the most dubious journals, given the extent to which the predators' marketing of their titles cleverly mimics that of established journals. Additionally, many of their articles are indexed on Google Scholar, a search engine playing a growing part in researchers' bibliographic research, despite the fact that Google does not publicly disclose how the search engine works or which databases it searches. Confusion is even greater among members of the general public, unable to find their way in the jungle of publications and liable to consider studies published in predatory journals credible and validated by the scientific community. This is how the old saw about vaccination as a cause of autism recently resurfaced (see Chapter 11) in the journal *Open Access Scientific Reports,* published by one of the biggest predators, the Indian company OMICS, despite the fact that it had been invalidated by every relevant study. Today, the fact that the study is scientifically inept does not prevent it from being deferentially cited by antivaccination groups, which, incidentally, are in close contact with its author, John B. Classen.

Toward Open Access for All

As worrisome as the development of predatory journals may be, it does not call into question the scientific publishing industry's major shift toward free access to articles. Like the music industry and, more recently, the film industry, scientific publishing has been hit head on by the growing demand that access to cultural products be free of charge and is looking for solutions to finance itself. However, unlike the music and film businesses, science is a field that is primarily funded by the public sector, throughout the world.

Consequently, the attitude of public authorities will determine the future of open-access scientific publishing, which currently looks very bright. While I have so far emphasized the dangers of predatory journals, these only account for a minority of the sector. Of the roughly ten thousand open-access scientific journals listed in the Directory of Open Access Journals and across all disciplines, two-thirds do not require any financial contribution from authors. These are primarily journals in the human and social sciences published by universities and research centers. "The concept of authors or their laboratories paying for their articles primarily developed in the biomedical field, because it is far better funded, particularly through its contracts with the private sector. Publishers realized there was a market there," confirms Maisonneuve.

The expansion of open access is also taking place in the experimental sciences. The German nonprofit research organization Max Planck Gesellschaft uses its own capital to fund open-access journals and asks its researchers to publish in them. In the United Kingdom, the Wellcome Trust, a leading foundation for the funding of biomedical research, requires that the researchers it supports publish in open-access journals and even launched its own open-access publication, *eLife*.

Maisonneuve adds, "Most scientific institutions and sponsors of research, including the European Commission and the American Congress, now encourage researchers to publish in open-access journals, in the name of the public's right to know the results of the research it funds through its taxes. In the end, only a few subscription journals will survive, owing to their historical prestige."

13

Beyond Denial

Nearly forty years ago, the American journalists William J. Broad and Nicholas Wade published *Betrayers of the Truth: Fraud and Deceit in the Halls of Science*. Their book was the first to focus on scientific fraud; it remains one of the few to have done so. Looking back at it today provides an assessment of the evolution of responses to fraud in the last three decades. While fraud is no longer denied the way it was then, the scientific community is still powerless to halt its progress.

Most of *Betrayers of the Truth* consists of a well-documented description of several cases of scientific fraud that took place in the United States in the 1970s (see Chapter 1). It could have been written yesterday. However, contemporary readers will be less surprised than those of the past to read Broad and Wade's long theoretical arguments aimed at convincing them that science is not a purely rational activity—that it is also shaped by financial, political, institutional, and ideological interests, as well as the researcher's temperament. Doesn't everyone know that by now? It has been well established by countless studies in the sociology of science, regardless of the branch considered. But these studies were only in their infancy when Broad and Wade wrote the book whose title so aptly expresses its accusatory and somewhat disappointed tone. Yet despite the antiestablishment fervor that had taken hold of part of the scientific community in the 1960s and 1970s, scientists seemed overall to be convinced that they were engaged in a disinterested activity that consisted in searching for the truth.

Betrayers of the Truth opens with an account of the statement made by the chemist Philip Handler, then president of the National Academy of Sciences, before a congressional investigations subcommittee in 1981:

> Instead of opening with the ritual profession of thanks for being asked to appear before the committee, Handler at once announced that it gave him "little pleasure and satisfaction" to testify on the subject of scientific fraud. The problem had been "grossly exaggerated" by the press, he said—as plain a way as any of telling the committee it was wasting its time. Scientific fraud happens rarely, and when it does, Handler declared, "it occurs in a system that operates in an effective, democratic and self-correcting mode" that makes detection inevitable. His underlying message came over loud and clear: fraud is a non-problem, the existing mechanisms of science deal with it perfectly adequately, and Congress should mind its own business.[1]

Donald Fredrickson, who was then the director of the NIH, was equally cutting in his testimony to the investigations subcommittee. According to Fredrickson, the very idea of committing fraud shook scientists "to [their] very core"; if they were caught, the penalties would be equivalent to "excommunication." Like Handler, Fredrickson recognized that there were no formal procedures in place to deal with cases of fraud. Anyway, he stated, these were more than marginal. But he went further by claiming that procedures of this sort were not desirable, because by definition they would require supervision from outside the scientific world to be impartial. Fredrickson stated he found it frightening to imagine laypeople trying to meddle with what took place in laboratories, for they could not "understand the requirements of the scientific method or the fact that its own correctives are in place."[2]

First Measures

Today, statements like those delivered by Handler and Fredrickson before the congressional investigations subcommittee would be unthinkable. The positions have been reversed. Thirty years ago, it was rare to find someone who admitted that fraud was frequent in laboratories. Today, it is rare to find someone who denies it. Scientific institutions—universities, research organizations, learned societies, specialized publications—all admit that the problem of scientific fraud exists. And that it is serious. Indeed, the congressional hearings held in 1981, in which future vice president Al Gore was a driving force, led to a legislative upheaval, first in the United States, then in Europe.

In 1985 Congress adopted a law stipulating that research institutions receiving public funding must put in place administrative procedures to resolve cases of fraud in their midst. This measure did not have much effect. As we have seen, the American scientific community was reluctant to allow the federal government to interfere in its affairs. Three years after the law was adopted, few American universities had complied with it. As for the NIH, it had only created one and a half positions to deal with scientific misconduct among its tens of thousands of researchers. But pressure from Congress became more and more insistent. Dennis Eckart, a representative from Ohio, made an explicit connection between the scientific community's reluctance to look at its failings and other cases of corruption and embezzlement that left their mark on the United States in the 1980s. He told Congress, "What is at issue here, much in the same way [as] within the defense industry, over at NASA, over at the accounting profession, over at savings and loans, are the adequacy of safeguards which will give you and me and the public confidence [that] waste, fraud, abuse or misconduct do not occur with taxpayers' dollars."[3]

Faced with growing political pressure, the major American scientific institutions decided to take action. In 1989 the National Science Foundation, the principal agency for funding American research, created an internal Office of Inspector General responsible for investigating and potentially sanctioning allegations of fraud. The same year, the NIH created its Office of Scientific Integrity. For the first time anywhere in the world, independent authorities were tasked with investigating cases of breaches of scientific integrity (which would soon be described as the fabrication, falsification, and plagiarism of data, as detailed in Chapter 1), which had previously been left to the laboratories' initiative. The Office of Scientific Integrity was provided with a regulatory enforcer, the Office of Scientific Integrity Review, which had the power to ban researchers found guilty of research misconduct from receiving NIH funding. In 1992 the two institutions merged into a single organization, the Office of Research Integrity (ORI). Its guidelines called for it to act within the framework of federal laws, particularly those that strictly regulate the use of public funds. From a legal perspective, scientific fraud was considered a form of misappropriation of funds, but without personal benefit.

Though I will return to the ORI in Chapter 14 to question some of its methods, at least in its early stages, its impact has been far from insignificant, with around 450 cases examined since it was founded. Notably thanks to its public awareness programs, the ORI's principal effect has been to make the scientific community less tolerant of fraud and to encourage scientists who witness cases of fraud to report them. It also has led to severe sanctions of the most flagrant cases of misconduct. While William Summerlin was merely considered overworked and in need of rest in 1974 (see Chapter 1), his unethical behavior would not go unpunished today. In fact, two researchers recently convicted of particularly severe fraud were sentenced to prison for embezzling public funds. Many researchers sanctioned by the ORI leave the scientific community.[4] Those who stay will have to deal with their tarnished reputations, difficulty getting published, and being denied access to NIH funding (which is part of the sanction). This does not only apply to those found guilty of fraud: any researcher who is made to retract an article will face the same problems. By analyzing 468 biomedical articles coauthored by 1,261 researchers and retracted from the biomedical literature between 1996 and 2006, the Quebecois researcher Philippe Mongeon showed that more than a third of these authors stopped publishing in the first five years following the retraction, and that the remaining two-thirds published less and were cited less often.[5] This relative blacklisting is far more acute in cases of fraud than of good-faith errors and has a greater impact on first authors (who are considered to have done the experimental work) than on last authors (considered to have overseen it).

From ORI to Europe

Most European nations now also have institutions specializing in the prevention and investigation of fraud. These were set up by individual countries rather than through a concerted action across Europe (the European Commission, already widely criticized for its cumbersome bureaucracy, avoided tackling the issue). As is often the case with issues of public morals, the Nordic countries were pioneers on the question. In 1992 Denmark was the first to create a "committee to combat scientific dishonesty." Remarkably, this initiative was taken despite the fact that there had been no cases of fraud in the country. Chaired by a judge, the committee follows the American approach of applying the law of the land to scientific fraud, particularly those laws concerning the use of public funds. Norway and Switzerland follow the same approach.

In Germany it took the resounding Herrmann case in 1997 (see Chapter 10) to push administrative authorities in the sciences to take action. In 1999 the Deutsche Forschungsgemeinschaft (DFG), the principal German agency for research funding, created an ombudsman office responsible for dealing with accusations of fraud. The German approach has a different philosophy from those in the United States and Denmark: it relies not on the law but rather on a codification internal to the DFG, which defines breaches of scientific integrity as well as investigation procedures. It also spells out a range of potential sanctions, of which there are six: reprimand, bans on submitting research projects for a period ranging from one to eight years, reimbursement of the funds collected, retraction of an article or publication of an erratum, bans on being an assessor of projects submitted to the DFG, and bans on standing for election to the agency's thematic committees. Like Germany, Finland and Poland have adopted systems based on a code internal to the national scientific community.

As for the United Kingdom, it began developing an even more decentralized system in 1998. Each of seven research councils dealing with specific disciplinary fields adopted its own code. Some, such as the Medical Research Council, were inspired by the German model. Others have adopted far less strict codes. Also in the United Kingdom, several publishers of biomedical scientific journals founded the Committee on Publication Ethics, notably in order to provide guidelines in cases in which fraud and plagiarism are suspected in an article submitted or already published.

As a result of these national efforts, the Council of the European Union adopted a declaration for research integrity at its meeting on December 1, 2015. This declaration is particularly laudable for outlining the harmful consequences of fraud in four different sectors:

a. Individuals and society: false results or unsafe R&I [research and innovation] products or processes may be released or may become public and widely accepted by the community or by other scientists with serious consequences, including hampering scientific progress;

  b. Public policies: unreliable data or untrustworthy advice may
     lead to poor policy-making;
  c. Public institutions: institutional abilities to foster and pro-
     mote research in a competent and responsible manner can
     be undermined;
  d. Public trust: research misconduct and the misuse of public
     funds can lead to the disruption of public confidence and
     support in science, and thereby endanger the sustainability
     of R&I funding.[6]

During the 1990s and 2000s, most scientifically developed nations put in place institutional systems to combat scientific fraud (I
discuss the specific case of France in Chapter 14). This diversity of
national approaches fostered a need to share experience, leading to
international conferences on research integrity in Lisbon (2007),
Singapore (2010), Montreal (2013), and Rio de Janeiro (2015).
A statement issued by the Singapore conference is reproduced in
this volume's Appendix. While it displays all the shortcomings of
international institutional doublespeak, it has the dual merit of reit-
erating sound principles and being able to serve as a consensual
foundation for countries seeking to put in place a policy for combatting scientific fraud.

Institutional Impotence

These unique bodies charged with handling allegations of fraud are
undoubtedly useful, whether operating on the scale of a country or
that of an institution. They provide researchers who witness breaches
of scientific integrity in their laboratories with the means to report
them. They offer written, standardized procedures to deal with scientific misconduct, which counters the scientific world's spontaneous
tendency to handle these problems through back-alley arrangements.
Finally, they encourage a commitment to scientific integrity on the

part of educational and training programs, particularly those for doctoral students and young researchers.

However, only a shockingly tiny fraction of cases of fraud are brought before these bodies. On average, the ORI has handled twenty-four cases of fraud per year for the last two decades, which is very few. The heads of the ORI are keenly aware of this. Three of them initiated an inquiry into the matter by sending a questionnaire to approximately 4,000 researchers funded by the NIH. They received 192 replies detailing 201 breaches of scientific integrity observed during the period from 2002 to 2005 (two-thirds were cases of fabrication and falsification, the other third plagiarism). In other words, three per researcher and per year. If this is extrapolated to the 155,000 scientists with the NIH, this rate would indicate 2,325 breaches of scientific integrity per year, or one hundred times more than the number of cases annually handled by the NIH.[7]

As for researchers, they are often hesitant to report perpetrators of fraud. In the ORI study just mentioned, only 58 percent of breaches of scientific integrity observed in NIH laboratories were reported to the head of the lab or the department.

The director of the ORI wrote,

> On an individual level, many reasons for under-reporting are easy to understand b ecause they involve motivations we might all have experienced. For example, one does not want to accuse falsely. One may also fear that reporting would take time away from research, or have concerns and fears about possible retaliation. One may assume someone else will or should report it. Or one may have sympathy towards a researcher, and might think "it's not too bad," it can be sorted out without a career-damaging investigation. Reporting also necessitates confidence that the issue will be examined carefully and thoroughly.[8]

The question of confidence in the institutions responsible for preventing and tracking fraud is a central one. A team of American researchers recently sent NIH scientists a questionnaire asking whether they had been witness to breaches of scientific integrity and, if so, what they had done. They received 2,599 responses. The fact that 85 percent of answers to the former question were positive can hardly be considered evidence that scientific misconduct is so predominant. Clearly, more researchers who witnessed such problems chose to respond to the questionnaire. Of the two-thirds of the respondents who chose to report the misconduct they witnessed, only 16 percent turned to their hierarchy or to their institution's specialized body. The others all opted for an informal approach, though the problem observed was only corrected in 28 percent of these cases.[9] A former researcher with the Pasteur Institute confirms that "the most frequent approach is to try to resolve the issue internally, discreetly, by using your connections: for instance, by talking to the doctoral program and ensuring that a researcher who lets his doctoral students beautify their data no longer has any students."

But to do that, you need to have connections. This is the case for laboratory directors and university professors. If determined to do so, they know how to discreetly remove someone who has committed fraud from the scientific community. But what happens when a young researcher, a postdoc, a technician, or an engineer on an insecure contract suspects a breach of scientific integrity? As they are responsible for the bulk of experimental work, these are the people most likely to witness potential fraud. Who would they report it to? Would they risk their careers or even their jobs to do so? The biophysicist Lucienne Letellier, a member of the CNRS Ethics Committee, openly recognizes that it is difficult for a young researcher to report suspected fraud:

Who do you tell when you find yourself dealing with colleagues who are little inclined to call their work into question? How do you talk about it? Who are your interlocutors in the institutions concerned? And if you do report someone, what protections are in place for you? Without clearly identified interlocutors, and under the effect of "friendly" pressure, you'll be tempted to give up on your initiative . . . and even if you did persevere, the mistrust and hostility of your peers and supervisors would stick to you. Ultimately, even if the transgression is recognized, it isn't necessarily the accused who will be forced to leave the laboratory, but the accuser![10]

This is a real danger, as can be seen from the sad case of Young-Joon Ryu, a former collaborator of Woo-Suk Hwang (see Chapter 2) at the University of Seoul. This researcher's career was shattered because in 2005 he informed the press of his former boss's countless acts of fraud. Since South Korea lacks an institution specialized in dealing with breaches of scientific integrity, Ryu did not have many other options to take action. Instead of being thanked for this brave initiative that notably—and happily—prevented Hwang from undertaking clinical trials on human beings, Ryu was virulently attacked in his country for having caused the fall of a national hero. Fired from his research institute, he struggled for two years to find a new laboratory.

The biologist Joseph Thomas faced an equally difficult situation after he contributed to making public a case of fraud at the institution that employed him, Duke University. Thomas sued Duke by invoking the False Claims Act (also known as the Lincoln Law), a federal law allowing individuals to file actions on behalf of the government against private entities that have misappropriated public funds. In the statement setting out his complaint, Thomas accused the university of "intentionally concealing the full extent of the scientific fraud" committed in its biology department. According to Thomas, these false results were used in dozens of applications for federal funding, which brought in a total of $200 million over nine years. According to the Lincoln Law, as a private entity the university could be sentenced to reimburse up to three times this amount if it is declared guilty. The stakes are high, and as I write these lines the legal marathon has only begun. Thomas's professional situation will remain terribly uncertain throughout the legal proceedings.

Though all the bodies specializing in combatting fraud guarantee that they will protect the privacy of those who contact them, the fear of being seen as an informer is powerful in a small community where everyone knows each other. Yet the Singapore Statement on Research Integrity (reproduced in the Appendix) states, "Researchers should report to the appropriate authorities any suspected research misconduct, including fabrication, falsification or plagiarism, and other irresponsible research practices that undermine the trustworthiness of research, such as carelessness, improperly listing authors, failing to report conflicting data, or the use of misleading analytical methods." But so long as rules for the legal protection of whistleblowers remain hazy, or even nonexistent, it is to be feared that few scientists will do their civic duty by reporting acts of fraud when they witness them.

Taking Justice into Your Own Hands?

As useful, novel, and laudable as they may be, institutional attempts to combat fraud seem to be reaching their limit. Sandra Titus, the director of health sciences at ORI, recognizes that, "despite attention to research misconduct and other issues of research integrity, efforts to promote responsible behavior remain ineffective."[11] According to her, the only effective measure would be to implement a kind of collective responsibility by hitting researchers in their pocketbooks. She proposes to prioritize funding to institutes that have put in place the strictest provisions to combat breaches of scientific integrity, in order to set in motion a collective virtuous dynamic, and conversely to encourage reporting breaches of integrity by penalizing the entirety of an institute where they have occurred.

The slowness and limits of institutional action probably explain why many researchers who have witnessed fraud or have suspicions about their colleagues' work make them public without going through the institutions. Since 2010 the enigmatic Clare Francis (a pseudonym probably used by a collective of researchers) has specialized in alerting the editors of molecular and cell biology journals of duplications, inversions, and retouched images in their articles. While a few journals, such as the *Journal of Cell Biology,* have retracted articles after receiving information from Francis, most editors refuse to consider these anonymous warnings, despite the fact that the Committee on Publication Ethics recommends taking them seriously.

It must be said that editors receive hundreds of these tips a year, and that it is nearly impossible for them to determine whether they are sent by researchers personally committed to integrity, dishonest competitors trying to get rid of their rivals, or pranksters. For similar reasons, journal editors nearly always refuse to take into account comments critical of their articles on the website PubPeer, which offers users the opportunity to anonymously criticize studies published in the biomedical field. Here, the author of the article is automatically notified of these critiques and can respond. While long relatively obscure, PubPeer achieved a sudden notoriety by publishing a series of comments extremely critical of the CNRS biologist Olivier Voinnet's method of conducting his experiments (see Chapter 14). The principal bibliographic database in the life sciences, PubMed, now offers links to the critical comments on PubPeer. The site's newfound fame has also led its creator to reveal his identity: Brandon Stell, an American researcher in neuroscience based at a CNRS laboratory in Paris.

14

Scientific Crime

As in the United States and Germany, it was media coverage of a resounding accusation of fraud in a biology lab that finally pushed research institutions in France to take action. The story deserves to be told in detail, for it reveals the central role political interventions long played in dealing with allegations of fraud in France.

In 1994 Inserm opened a new laboratory, Unit 391, at the pharmacy school in Rennes. The lab focused on studying the metabolism of lipids and was headed by a young Belgian scientist, Bernard Bihain, who had just identified a liver protein involved in fat degradation. This discovery suggested the possibility of a new treatment for obesity and was of significant interest to the biotechnology company Genset. Genset and Inserm made a collaboration agreement to study what was then occasionally hastily referred to as the obesity gene. But in 1997, some of Bihain's coworkers began accusing him of fraud. One of his collaborators reported, "I had done an experiment that hadn't worked. I told Bihain, who told me not to worry, that we would do it again. But he still used the results, though they were unusable. I got worried that down the road I would be accused of having distorted the data myself."[1]

The laboratory was affiliated with the University of Rennes 1. The Office of the Chancellor was notified of the allegations and took the case seriously. But it was Bihain who requested that an investigation committee be appointed by Bernard Bigot, then director general for research and technology at France's Ministry of Research. After collecting twenty-four witness statements, the committee turned in its report on October 28, 1997: "The committee unanimously . . . judged that the accusations the director of Unit 391 has been charged with are serious and corroborating and that he will need to respond to them. A thorough analysis of the accusations and responses will require convening scientists with the expertise specific to the field of research under consideration."

Despite these indisputably harsh conclusions, Bigot chose not to follow up on the committee's suggestion to take the investigation further. Acting on his sole authority, he ruled that the case was closed. The administration of the University of Rennes 1 bitterly commented that "it [was] highly surprising that the director general for research and technology replaced the experts he had appointed in order to personally complete an investigation for which he had none of the required scientific expertise."[2]

The case suddenly came back to life in February 1998, when two of Bihain's colleagues asked to remove their names from the author list of an article he had cowritten, in order to avoid being associated with what they described as a "false interpretation" of experimental results. Things became more acrimonious when a report by the University of Rennes 1's Committee on Hygiene and Security severely criticized Bihain's methods of running his laboratory, accusing him of authoritarianism, psychological harassment, and negligent handling of toxic products. Without waiting for a second investigation committee's findings on the accusations of fraud, which would never be filed, Inserm chose to dissolve Bihain's Unit 391. He energetically contested all the accusations leveled against him, then left for the United States, where he had begun his career.

In 1999 the Bihain fiasco led Inserm's senior management to create an internal Delegation for Scientific Integrity responsible for carrying out investigations of suspected cases of fraud. This was the first time a French scientific institution had dealt with the issue directly. This initiative created a precedent followed by the Pasteur Institute, which launched its own Committee for Deontological Supervision and Conciliation (CVDC) three years later. It is no coincidence that the first French organizations to put in place institutional means to deal with scientific misconduct were two institutions involved in biomedical research, the field most affected by fraud. After lagging behind for many years, in 2011 the CNRS created a position for a mediator charged with collecting all complaints and steering them toward the appropriate department. In 2014 the CNRS Ethics Committee published a useful guide entitled *Promouvoir une recherche intègre et responsable* (Promoting honest and responsible research). Finally, the CNRS's board of directors adopted a procedure for the treatment of accusations of fraud. In the meantime, INRA also instituted a policy for promoting scientific integrity by equipping itself with a delegate for deontology.

And in Practice?

An op-ed piece published in *Le Monde* in the fall of 2014 suggests that the French scientific community is no longer in denial. In a statement that would have been unthinkable a decade ago, some twenty French researchers, most in biomedicine, declared, "There are numerous documented cases of fraud, in every scientific discipline and in most countries. . . . Beyond proven frauds, one can list numerous breaches of scientific integrity, which are often tolerated, or even accepted in the research field: selecting experimental results based on preconceived ideas, dividing data to increase the number of publications, omitting previously published results, using inappropriate statistical tests, etc. While apparently minor, these transgressions surreptitiously and passively accepted in research circles must be taken seriously for they open the door to more serious misconduct."[3]

One should also note the creation of a working group dedicated to scientific integrity by the Mouvement universel de la responsabilité scientifique (Universal Movement for Scientific Responsibility). This working group has organized several symposiums and public initiatives where leading researchers engaged in conversation with political leaders and members of parliament.

Now that the problem of fraud is recognized, measures have been taken to deal with it: with the notable exception of the CEA, the principal French research organizations now have bodies responsible for combatting fraud, though these have a variety of appellations. Most were founded too recently for their impact to be assessed, with the exception of the most senior initiative, Inserm's Délégation à l'intégrité scientifique (DIS, Delegation for Scientific Integrity), which has existed for fifteen years. Yet Martine Bungener, who headed the DIS from 1999 to 2008, admits, "It must be recognized that while the creation of the DIS allowed us to effectively resolve many problems, it did not succeed in checking the increase in breaches of scientific integrity."

The DIS handles approximately fifteen cases per year (this figure has remained stable for a decade), while Inserm is an organization that employs about ten thousand individuals, half of whom are researchers. In a perfect example of the difference between American and European notions of breaches of scientific integrity, its scope of referrals is much larger than that of the ORI. Unlike the American institution, the DIS can be referred to for conflicts arising over the authorship of an article: this is a particularly sensitive issue in biology and medicine, where an article's first author is considered to have done the work, the last author to have headed the research, and the others to have contributed in various ways. For researchers, both credit as an article's author and the order in which one is listed are crucial issues since they define the future recognition of each individual's contribution to a research project. These conflicts over authorship make up the vast majority of cases handled by the DIS, with suspected cases of fraud only reported an average of three times per year.

Outside scientific circles, conflicts between coauthors might seem petty. One might be surprised that the DIS devotes most of its efforts to attempting to resolve them. In fact, these disagreements over authorship are indicative of a wide variety of larger conflicts. As Bungener puts it,

> I have observed that you rarely find breaches of scientific integrity in laboratories where there are no conflicts. If I were to give one piece of advice to someone wanting to commit fraud, it would be to avoid having any conflicts! If you do, the fraud will always wind up being reported. What strikes me is that plaintiffs, who present themselves as victims by referring a case to the delegation, are most often revealed to have also committed a breach of scientific integrity at some point or other. There are no good guys and bad guys, but a world of gray, in which all the protagonists in the conflict are more or less involved.

Persistent Taboos

France still falls short of the initiatives to combat scientific fraud found in northern Europe and the United States. The *Le Monde* op-ed by a group of researchers noted, "Too often these cases of misconduct are discreetly handled, sanctioned or hushed up." The old habit of sweeping things under the rug dies hard. In 2007 the Ministry of Research announced that it was appointing CNRS executive Jean-Pierre Alix to write a report on scientific fraud. The letter defining Alix's mission had the merit of being frank. It declared, "In France, the treatment of scientific fraud is essentially limited to sanctioning obvious cases when they surface," and continued with the understated admission that "this situation involves an unevaluated risk and is not the most favorable for maintaining a very high level of scientific integrity." Alix's report was filed in 2010 but never made public. Despite my repeated requests, the ministry did not choose to share it with me. Yet it contained highly interesting information.

For instance, the report's author investigated the thirty public research organizations in France. A third of them, including the CEA, stated they were not aware of any problems of fraud within their ranks. "One can question the value of this information," the report's author discreetly noted, before underlining that it was impossible to measure the extent of scientific fraud in France because of the lack of appropriate tools: "Institutions also frequently adopt the approach of hiding fraud. In this case, the driving force is the fear of losing a positive image built up in the community and among the public. So, people avoid talking about it. . . . Contrary to what these practices suggest, when asked individually, colleagues all know of several cases of fraud or misconduct, which implies that they are quite widespread, though this does not amount to a rigorous verification." The report also suggested a plan of action, the principal measures of which are discussed later in this chapter. Four years later, I asked Alix what effect he thought his report had had. "None," he replied in a disillusioned tone, before specifying, "at least not in terms of ministerial action."

Another reason to doubt the intensity of French efforts to combat scientific fraud is the Balkanization of the French scientific system, a paradoxical trend in such a traditionally centralized nation. In his report, Alix wrote, "The French system [for combatting fraud] is not formalized on a national level. . . . Often the treatment [of fraud] is kept secret and sanctions are taken with a great deal of discretion. When the treatment is codified, it varies from one place to another. Finally, many decentralized scientific institutions do not have any known regulations."

Of the roughly seventy universities in France, which are increasingly supposed to be at the heart of the system of national research, those that have provided themselves with a delegate for scientific integrity can be counted on a single hand. Programs to train doctoral students in scientific integrity remain marginal. When these programs do exist, such as at the University of Paris 5, they are coordinated by a private organization (L'Atelier des jours à venir) rather than the university—a clear sign that the academy is uncomfortable dealing with these issues. The French academic world continues to tolerate practices incomprehensible abroad, as demonstrated by a colloquium entitled "Ethical Approaches to Plagiarism," held at the University of Paris 8 in May 2014. Several participants were professors who were themselves plagiarists or had covered for their doctoral students' plagiarism. A few months later, the president of the Free University of Brussels resigned after using an unattributed part of a speech by Jacques Chirac in one of his own addresses (which he had not written himself), despite the fact that this instance of plagiarism was in no way damaging to the rigor of scientific knowledge. The French academic world would do well to learn from this example of high deontological standards.

Even the measures French public research organizations have put in place for dealing with allegations of fraud do not appear particularly transparent. For instance, an analysis of the procedure for DIS internal investigations at Inserm reveals that an investigation can at any moment be short-circuited by the organization's director general. The process for responding to a reported breach of scientific integrity in any of the institute's laboratories follows three stages (which are basically the same in all the other French organizations). The first is local treatment. But the director general must give his or her consent for the delegate for scientific integrity to be able to proceed with this preinvestigation with his or her local counterparts. The second stage is the "recourse to expertise." But the experts asked to provide their assessment of a presumed breach of scientific integrity are all appointed by the director general alone. Once the director general has been informed of their conclusions, he or she alone will decide "how to follow up on the case." As for the third stage, "in-depth investigation," the procedure is entirely at the initiative of the director general. According to the DIS's rules and regulations, it is up to the director general "to decide upon implementation of the subsequent modes of investigation, for the procedure might no longer remain confidential. At this stage, the procedure could indeed require an on-site investigation, with inspection of the laboratory notebooks and other documents, interviews with witnesses, etc."

It is understandable that these internal regulations display a concern for discretion. But it should also be noted that they do not do much to encourage transparency, that they allow Inserm's management to bury the case at any moment, and that they do not provide any measures for those accused to have the right to defend themselves.

An Academician Suspected

The case of the neurobiologist Henri Korn confirms that these are not mere suppositions on my part. Korn was bringing to a close a distinguished career as a professor at the Pasteur Institute and a member of the French Academy of Sciences when a violent debate erupted about the reliability of his findings on the neurophysiology of synapses, most of which dated from the 1980s. In 2002, with a truly foul atmosphere prevailing in Korn's laboratory, the administration of the Pasteur Institute was notified of serious statistical incongruities in the articles that had made his reputation. Late in 2003 the institute's director general reluctantly referred the case to the CVDC. The committee received two written documents suggesting that the figures published in some of Korn's articles were fallacious. One was a report by Nicole Ropert, a former colleague of Korn's whom he had dismissed from his laboratory at the very point when her research failed to confirm the theories that had made him famous. The other was a rigorous statistical analysis of the data in several of Korn's articles, prepared by the biologist Jacques Ninio, who does not work for the Pasteur Institute. The analysis demonstrated that

Korn's articles contained errors so serious that they needed to be retracted. Ninio demanded a written response to his criticisms, if not from Korn, then at least from the statistician who had analyzed the data in question. The CVDC refused, arguing that it needed to protect the confidentiality of its work. In fact, the committee cannot interview witnesses without the director general's authorization, which significantly limits its independence.

In 2006 the case grew uglier when the chairman of the CVDC received a letter from Korn's lawyer informing him that the procedure under way faced a "problem regarding essential proof": twenty years after the fact, Korn could not provide the raw experimental data that had led to his contested publications. Yet Korn's lawyer went even further. Knowing that the CVDC was planning to convene a committee of international experts to decide on the case, she reminded its chairman that this procedure "is not acceptable because it goes against the law of contradiction, stated in article 16 of the new code of civil procedure and confirmed by article 6 of the European Convention on Human Rights." The argument was irrefutable from a legal standpoint but highly debatable from the perspective of scientific tradition. One year after this formal notice was delivered, the Pasteur Institute's new director closed the investigation. It had ultimately served no purpose in elucidating whether Korn's articles can be considered valid.

The fiasco convinced Ninio to make the case public by publishing an article drawn from his report to the CVDC in a peer-reviewed journal. In 2007 the *Journal of Neurophysiology* published Ninio's rigorous statistical analysis of the charts in Korn's articles, in which Ninio noted "severe anomalies" and described results that were "almost miraculous."[4] Korn and his collaborators published a response in the same journal, but it was not particularly satisfying. Serious doubts remain regarding the scientific integrity of this neurobiologist with the Pasteur Institute. The data that made his reputation was obtained at a time when the analysis of recordings of neuronal activity was done by hand, or rather by eye, by observing the printed curves produced by the electrophysiological recording apparatus and measuring them with a ruler. Korn's problems started within his own laboratory, when this analysis began to be computerized. This removed any possibility of bias on the part of the analyst, who could no longer be tempted to choose only the data most favorable to his or her theory. This new method of analysis totally disproved Korn's previous results. But instead of duly noting this potential error due to the illusion of finding what one would like to find in experimental data, this researcher known for his extreme authoritarianism unceremoniously fired those of his coworkers who had established the use of computerized analysis in his laboratory. Today these former coworkers state they have little doubt that Korn was not impartial in analyzing his data.

Another Academic Discredited?

In September 2014 PubPeer, a website for discussion of scientific articles, published a series of comments pointing out anomalies in the figures presenting the data of several articles coauthored by the biologist Olivier Voinnet. At the time, this French researcher was known worldwide for his participation in the discovery of RNA interference, a process that regulates gene expression and provides plants with a kind of immunity. This discovery promised Voinnet a dazzling career. Recruited to work as a scientist for the CNRS a year after submitting his dissertation, he founded a laboratory at the Institute of Plant Molecular Biology in Strasbourg and continued to study RNA interference. In 2010 the CNRS sent him on a temporary assignment to the Swiss Federal Institute of Technology (ETH) in Zurich, where he headed a team of thirty people. Recognition came fast: he was awarded the CNRS Silver Medal in 2008 and the European Molecular Biology Organization Gold Medal in 2009. In November 2014, at the age of forty-three, he was invited to join the French Academy of Sciences.

The anomalies reported in the comments on PubPeer relate to figures depicting results of gel electrophoresis, a technique for separating proteins or nucleic acids, which appear as dark bands in a clear gel. The comments pointed out that in Voinnet's "control" figures these bands are often identical from one figure to the next, or reversed in a mirror image, and that they sometimes seem to have been cut out and repositioned. The suspicion of fraud was considered serious enough for the CNRS and ETH to initiate internal investigations into what appeared to be the first major case of fraud involving a French researcher since the 1990s.

In June 2015 the two investigative commissions delivered their conclusions. The CNRS commission's report was not made public; its findings were only commented on in a brief communiqué stating that there were proven manipulations of figures in thirteen articles coauthored by Voinnet. Judging that these "occurrences are not due to a simple succession of errors but are the results of bad practices" and that they constitute "serious breaches of the principle of integrity in scientific research," the CNRS dismissed Voinnet for two years "starting from the decision to terminate his assignment" to ETH. This was the heaviest sanction ever imposed in France for a case of scientific fraud.

In a detailed, twenty-two-page public report that included Voinnet's responses, ETH's investigative commission came to a more qualified conclusion. Of the thirty-two offending articles, it considered twenty problematic due to image retouching, which the researcher acknowledged he had performed. Five articles were retouched to the point of requiring retractions—that is, being removed from the scientific literature. Others required a corrective, since the data they reported was considered valid after careful examination by the commission.

The report notes that "the tragedy is that the experiments that were (mis)reported in the papers investigated by the commission were actually conducted and, as based on the data inspection by the commission, executed with care." The ETH commission therefore dismissed the accusation of fabrication of data (that is, it was not simply invented). According to the commission, the acts held against Voinnet—and which he recognized—were violations of the organization's internal code for proper scientific practices, notably a failure in his duties to be "vigilant regarding the figures published and [exercise] oversight as a team leader." ETH limited its sanction to issuing Voinnet a warning and decreeing that going forward in his work for ETH, Voinnet would have to be "accompanied by an external specialist to implement the measures necessary to improve his work conduct."

Because it unfolded in both France and Switzerland, the Voinnet affair provides a useful example of the diversity of national approaches to treating allegations of fraud. In France, which was long reluctant to deal with these issues, a choice seems to have been made to "make an example" by issuing a heavy sanction, but at the conclusion of an opaque procedure. In Switzerland, the exact opposite choice was made.

In the fall of 2018, following a new investigation, while the CNRS found Voinnet guilty of insufficiently leading his teams, he was not found guilty of fraud. The manipulated images initially published on PubPeer were in fact due to mistakes made by Voinnet's former collaborator, today an administrator of the South Province of New Caledonia. Voinnet's scientific reputation is restored. But he still finds the CNRS's initial action outrageously harsh.

Reforms

Some might chuckle at the obsession with litigation that seems to characterize public life in the United States. But one has to recognize that it implies a level of transparency in the investigation of suspected cases of breaches of scientific integrity that France is far from achieving, as shown by the Bihain, Korn, and Voinnet cases. All of the documents governing the activity of the ORI are available online, down to the slightest detail. It should also be noted that the ORI's director only reports to the secretary of health, who oversees the NIH, and not to the NIH's director. Similarly, the office of the German ombudsman is independent of the funding agency's management. Yet in France, as we have just seen, those overseeing scientific integrity at research organizations are subject to the authority of these organizations' directors.

And the websites for the CNRS, Inserm, the Pasteur Institute, and the INRA do not feature any documents describing the procedures for investigating suspected fraud. These documents exist, of course, as evidenced by the fact that I was able to access some of them, but they are not made public. Most egregiously, they are not accessible to researchers whose work is being investigated and who might well feel as if they are caught in a frightening, Kafkaesque mechanism they know nothing about.

According to Alix, a dedicated policy for promoting scientific integrity in France would need to be based on four essential elements. The first would be a national charter for scientific integrity, based on the Singapore Statement on Research Integrity adopted in 2010 (see Appendix). The second would be to include scientific integrity in all university curriculums. Alix has an idea of what such a program should look like: "One can imagine a consciousness-raising program in the very first years, those when a student must give up high school habits to gradually become a producer of new knowledge. The question of plagiarism would loom large at this stage. Then more in-depth training in doctoral schools, since many future doctors envisage a career in research." The third is to codify the practices for investigation of allegations of fraud within each institute. "We need explicit and public procedures, if possible with a table of standardized sanctions, protecting whistleblowers and providing for appeals," explains Alix. The fourth and last element takes into account a French particularity, which is that most researchers are civil servants. Alix concludes by saying that this would require "revising and modernizing the 1934 law on the deontology of civil servants so that it can serve as a legal basis for potential legal actions in cases of known breaches of integrity." Four pertinent proposals waiting for the political will to implement them.

A Global View

In 1942 the American sociologist Robert Merton published his famous analysis of the scientific community's ethical norms, noting that there was a "virtual absence of fraud in the annals of science."[5] According to Merton, the norms in place made fraud doubly impossible: on the one hand because fallacious results would be detected thanks to the "organized skepticism" that made the critique of any new finding a collective imperative, and on the other because the "norm of disinterestedness" implied that it is not in any researcher's personal interest to commit fraud, since research is carried out in the sole interest of knowledge, rather than personal gain.

A decade later, Merton had changed his mind. In a 1957 address to the annual congress of the American Sociological Society, Merton concluded that, "the culture of science is, in this measure, pathogenic."[6] In particular, he blamed "the norm of originality," by which the first person to carry out an experiment or formulate a theory is the only one to benefit from it, following the "winner take all" principle discussed in Chapter 6.

Commenting on various historical examples of data cooking, theft, and plagiarism, Merton explained, "The great cultural emphasis upon recognition for original discovery can lead by gradation from these rare practices of outright fraud to more frequent practices just beyond the edge of acceptability, sometimes without the scientist's being aware that he has exceeded allowable limits."[7] Merton's new position came in the context of his research on deviance, which he showed could be due to the social structure itself, and particularly its value system. He wrote, "Great cultural emphasis upon the success-goal invites this mode of adaptation through the use of institutionally proscribed but often effective means of attaining at least the simulacrum of success—wealth and power. This response occurs when the individual has assimilated the cultural emphasis upon the goal without equally internalizing the institutional norms governing ways and means for its attainment."[8]

Similarly, to paraphrase Merton, the scientific community's extreme emphasis on the originality of a discovery incites researchers to use forbidden but often effective practices such as fraud, falsification, and plagiarism to attain at least a simulacrum of success: the publication of articles in the most prestigious journals.

Merton's approach to fraud as social deviance raises the question of whether fraud can be treated as a form of white-collar crime, like financial crime. As we saw in Chapter 11, scientific fraud can lead not only to the embezzlement of public funds but also to endangering human lives. This suggests that it might be better for fraud to be handled by the criminal courts rather than research institutes' internal commissions, whose limitations we have had occasion to observe. This tempting idea has been the subject of growing debate in the scientific community. Richard Smith, the former editor in chief of the *British Medical Journal,* sees three reasons to consider the worst cases of fraud as punishable under criminal law: "First, in a lot of cases, people have been given substantial grants to do honest research, so it really is no different from financial fraud or theft. Second, we have a whole criminal justice system that is in the business of gathering and weighing evidence—which universities and other employers of researchers are not very good at. And finally, science itself has failed to deal adequately with research misconduct."[9] Though Smith's arguments are convincing, I think there is a simple, practical reason that turning fraud into a criminal offense is not such a good idea: the few attempts to involve the police and the courts in resolving accusations of fraud have been fiascos.

A Nobel Prize Winner Accused of Fraud

The American David Baltimore, winner of the 1975 Nobel Prize in Physiology or Medicine, can bitterly testify to just such a situation.[10] In 1986 Baltimore was one of five coauthors of a study published in *Cell*. One of his coauthors was Thereza Imanishi-Kari, who headed a cellular immunology laboratory at MIT. A few months later, Margot O'Toole, a young postdoctoral researcher at the same laboratory, claimed that she could not reproduce the *Cell* article's results and brought the case before the relevant academic institutions. Commissions at Tufts University, which was preparing to hire Imanishi-Kari, and at MIT successively concluded that the article only had minor errors. According to these commissions, the nature of the controversy was purely scientific: it could be resolved by further research.

Yet the affair took on an entirely different scope when O'Toole decided to become a crusader for scientific integrity and to alert a variety of political figures at a time when Washington, DC, was very concerned with scientific fraud. John Dingell, a Democratic representative from Michigan, took a passionate interest in the case. He headed three congressional hearings on scientific misconduct, all of which were focused on Imanishi-Kari and far from impartial: in his zeal to denounce fraud, Dingell seemed as fervent as Senator Joseph McCarthy three decades earlier in his efforts to root out alleged communist infiltration. The NIH, which had just created the Office of Scientific Integrity (see Chapter 13), saw the case as an opportunity to prove its determination to combat fraud. The Secret Service, a federal agency tasked with fighting counterfeiters and financial fraud, seized Imanishi-Kari's laboratory notebooks. It later disclosed that its analysis showed data falsification. In 1991 the Office of Scientific Integrity concluded that Imanishi-Kari had committed fraud. Baltimore, who had steadfastly expressed his support for his colleague, was forced to resign from the presidency of Rockefeller

University in New York, a position he had stepped into just eighteen months earlier. Imanishi-Kari was fired from Tufts and barred from receiving research funds. Three years later, the ORI, which had replaced the Office of Scientific Integrity, confirmed this verdict.

Yet the investigation conducted by the NIH and Congress with the technical support of the Secret Service proved to be biased. At no point were Imanishi-Kari and Baltimore given access to the case files gathered against them, nor were they given the opportunity to respond to the accusations leveled at them other than on the defensive, since they were forced to prove their innocence from the outset. The Secret Service agents, who were more accustomed to hunting down counterfeit bills than assessing the results of experiments in cell biology, had made crude mistakes in their analyses. Or, to be more specific, they had applied the strict police rigor of their investigative methods to the study of laboratory notebooks, which are often kept with a degree of carelessness. At the time, it was common to update lab notebooks only on a weekly basis, in order to report the week's experiments. An experiment carried out on the 8th might therefore be dated the 12th: in the eyes of the Secret Service, this backdating was an infraction. Similarly, the agents were shocked to discover that an experiment image in the laboratory notebook had in fact been obtained by agglomerating the results of several manipulations: to biologists, this is a routine practice, but according to the Secret Service it was punishable under criminal law. Things have partially changed since then, but at the time, researchers considered the laboratory notebook a kind of aide-mémoire, practically a private journal, which they never would have imagined could one day be used as a piece of evidence in a court of law.

In 1996 an appeals panel cleared both researchers. This "bittersweet" exoneration, as Baltimore describes it, confirms my sense that it is dangerous to turn to the police to establish the legitimacy of an allegation of fraud. As Baltimore writes, "I believe that the 10-year controversy proved that congressional committees are no place to adjudicate scientific issues, and I am glad that there has never been another such case."[11] Baltimore also points out that if the Republicans had not taken the House and relegated his inquisitor, Dingell, to the minority, thus depriving him of much of his power, the case that now bears his name might have had a far different outcome.

Res Judicata

In the first part of this chapter I discussed the accusations of fraud made against the biologist Bernard Bihain in the late 1990s. I had left this affair with the closing of Bihain's Inserm laboratory at the Rennes pharmacy school. But the story does not end there. In the summer of 1999, the president of the University of Rennes notified the public prosecutor's office that there was serious suspicion that Bihain had modified a chart taken from various experiments conducted in his laboratory in several documents, including a publication, a patent submission, and an activity report. According to French law, this is known as "forgery and use of forgeries." A legal investigation was launched. In the summer of 2003, the Rennes high court ruled on the case. The ruling began by stating that according to article 441-1 of the penal code, the category of "forgery and use of forgeries" only applies to written documents liable to establish evidence of a right or fact that has legal consequences. It cannot therefore apply to scientific articles. Concerning the other documents in question, the court stated that the facts brought to its attention "were essentially a controversy of a scientific nature, arising between researchers disagreeing over calculations, interpretations of raw data, issues of standardization, methodological disagreements, etc. . . . It is not established that Bernard Bihain fraudulently altered the truth, whether in the quadrennial report or the patent submission. . . . It appears obvious that the dispute between Bernard Bihain and those accusing him is primarily a question of scientific rigor and not of forgery as it is defined in article 441-1 of the penal code." With this, the case was dismissed.

Like Baltimore, Bihain was cleared after years of suspicion; he briefly returned to Inserm before continuing his career in the private sector. But what really happened in that laboratory in Rennes from 1994 to 1998? Since the court did not rule on the question, and in the absence of any in-depth investigation, we will probably never know. And most significantly, it will be impossible to take a second look at the case, because of the legal principle of res judicata.

The issue of the relationship between the violation of common law and scientific fraud returned to the fore with the discovery of misconduct by the biologist Erin Potts-Kant of Duke University in North Carolina. This specialist in lung physiology was caught using the laboratory credit card to spend more than $25,000 for her own pleasure. Potts-Kant was fired, but her embezzlement led to the university's administration to question her intellectual probity—which proved to be deficient. Some of her experiments had been improperly conducted, or even invented, while a great deal of her data was fabricated, as was confirmed when researchers tried and failed to reproduce her results. More than fifteen articles by Potts-Kant were retracted. A common case of the misuse of company assets led to one of the most significant cases of scientific fraud in recent years.

More recently, the Italian biologist Enrico Bucci, the head of Bio-Digital Valley, the firm whose work demonstrated that one-tenth of the biomedical literature's images showing protein modifications in different human pathologies are falsified (see Chapter 5), reported to the Milan police in 2013 that the Neapolitan researcher Alfredo Fusco, a specialist in oncology, had significantly overindulged in image retouching in his publications. According to Bucci, "In Italy, there are no laws or regulations dealing with scientific fraud that are internal to universities and research organizations. Consequently, when misconduct is suspected the only solution is to refer the case to the police, which will investigate it by using other legal categories, such as the misappropriation of public funds."

The police investigation's findings are still pending, but eight articles by Fusco have already been retracted. Though the police pride themselves on being scientific, the Baltimore case is only one demonstration of the fact that they are often poorly equipped to analyze laboratory notebooks. And it is to be feared that police investigations will fail to prove the existence—or nonexistence—of falsifications in the work published by Fusco.

15

Slow Science

While thinking of fraud as scientific crime seems like a dead end, Robert Merton's key argument that social structures inherently produce deviance is an excellent framework for reflection. His approach is particularly valuable for underlining that it is delusional to believe that we can reduce the extent of fraud through either technical solutions (software for detecting plagiarism and image retouching, increased statistical control) or institutional ones (as outlined in Chapters 13 and 14). To tackle the problem at its root, we must modify the social structures of science.

These "fraudogenic" social structures can be divided into two categories: those that are internal to the scientific community, such as the publication system, and those that are (partially) external, notably the evaluation of academic researchers' work by their employers, universities, or research institutes. This chapter examines recent initiatives and proposals to reduce the extent of fraud and other breaches of scientific integrity in both categories.

Communalism

In looking at the norms governing the scientific community, Merton emphasized the importance of what he called "communism," often referred to today as "communalism" to avoid political connotations.

According to this norm, any discovery belongs to the entire scientific community. Outside of research and development for industry, there is no call for secrecy. Any allegedly new scientific knowledge must be made public, in full detail. While this norm is clearly not always followed in practice, it is certainly a founding principle of the ethics of science.

The internet has allowed communalism to reach a new scope. While researchers could once only share their results by turning them into articles—which, as we saw in Chapter 3, required them to rewrite the history of their research—today they can share their raw data, make public the various stages of the data processing, and ask their colleagues to comment on early drafts of their articles. Theoretical physicists were the first to take advantage of these new technological possibilities with the 1991 launch of arXiv, now a highly respected online data repository that has since expanded its coverage to certain areas of computer science and mathematics. The only requirement to post a study to arXiv is that it be done from an electronic address hosted by a scientific institution and, in some cases, that it be sponsored by a well-known researcher for a first publication. Today biology is slowly converting to this new way of circulating research results, which amounts to bypassing journals and their peer reviewers. Websites such as Dryad, Figshare, and Runmycode offer researchers the opportunity to post all of an article's data and methods, which allows anyone to reproduce their experiment (in theory) and, especially, to verify their analyses and calculations (in practice).

But let's not be naïve. Part of the reason theoretical physicists began publishing their raw data close to thirty years before biologists was because the economic stakes are far lower in their discipline. In the biomedical field, everyone closely guards data that can lead to financial windfalls through patents and contracts with private companies. But as we saw in Chapter 5, the pharmaceutical and biotechnological industry is far more concerned than the academic world with the difficulty of reproducing published results in the life sciences. Part of the problem is that publication is only a reconstructed, selective view of the research carried out (see Chapter 3). If raw data were available, it would become much simpler to understand the difficulties encountered in reproducing an experiment. If anyone could consult experimental results, whether colleagues or competitors, results would be more effectively controlled. It thus makes sense that a group of NIH experts who spent a year studying how to improve the reproducibility of results in the life sciences made the public sharing of raw data one of their principal recommendations.[1]

Except among specialists in theoretical physics, data sharing has not yet become the norm. Nonetheless, this practice, still unknown a decade ago, is spreading quickly in other disciplines. According to a survey by the scientific publisher Wiley, approximately half the researchers in the world share their raw data in some form, with large variations from one country to another.[2] The Germans share the most (55 percent) and the Chinese the least (36 percent). However, very few researchers (6 percent) use specialized sites such as Dryad, Figshare, and Runmycode to put their data online, preferring to use the option offered by most scientific journals of accompanying their articles with "supplementary materials" (which entails a certain treatment and selection of data) or publishing them on a personal web page. Equally interesting are the reasons respondents to the survey provided for not wanting to share their raw data. Unsurprisingly, issues of intellectual property and confidentiality came first, with 42 percent of responses, followed by the fear of having results stolen (26 percent). But an unexpected reason sneaked in between these two others: 36 percent of researchers argue that they do not share data because it is not "a funder requirement." A potential way to generalize the sharing of raw data would therefore be for it to be required by institutions (universities, research organizations, private firms, and so on).

The Virtues of Sharing

Putting raw data online would also make research's hidden face more visible—namely, the vast body of results that are never published because they are considered "negative" (see Chapter 4). The website ClinicalTrials.gov provided a first initiative in this direction by offering organizers of clinical trials a platform on which to post the reasons that led them to conduct a study from its very beginning. Despite the fact that several leading medical journals require trials to be listed on ClinicalTrials.gov, fewer than half of clinical studies are currently registered. Yet registering on the site has the clear advantage of forcing researchers to actually answer the question they were initially asking themselves when it comes time to publish their study, rather than reformulating it in order to obtain a positive answer through the all-too-common practice of HARKing (see Chapter 4).

In June 2013 an initiative entitled Restoring Invisible and Abandoned Trials called for the publication of any trial in the field of clinical research that had never been published because it did not seem to provide any new information.[3] Taking advantage of new freedom-of-information policies allowing the public to access data that private companies submitted to health authorities, the authors of this call announced that they had in their possession no fewer than 178,000 unpublished pages of internal pharmaceutical company reports on trials carried out with widely prescribed drugs. In their call published in the *British Medical Journal,* the authors challenged pharmaceutical companies to publish the results of these studies, failing which they would make the raw data public. This ultimatum has started to bear fruit, with the publication of unpublished clinical trials going back to the 1990s.

More generally, scientific publishing must fully deal with the consequences of the digital revolution. The scientific article, that well-told tale (see Chapter 3), has no future if the data that informed it is not uploaded to support it. Given the rate at which daily collection of data and e-signature certification are developing, the reliable old laboratory notebook will be a mere memory in a few decades. Even the anonymity of referees, a long-standing practice in scientific publishing, is increasingly less justified. It has become technically possible, if desired, to make public every stage on the tortuous path to the conclusions reported in a study. I have already mentioned arXiv, used in the fields of mathematics, physics, and computer science.

Climatologists are now also using these less competitive and more collaborative methods to share research results. Researchers specializing in the climate history of the earth can submit their manuscripts to the site for the journal *Climate of the Past*. Manuscripts are evaluated both by two peer reviewers selected by the editor and by anyone who wants to make comments. Every submission is ultimately published online, but the site indicates whether it was accepted by the "official" reviewers.

Can these highly communalist methods—to use Merton's term— be applied to extremely competitive fields with high economic stakes, such as biomedicine? There is no reason to doubt it, unless one were to doubt their scientific nature. Some journals in the field have made the leap, including those published by the PLoS group, which created a section on the Dryad data repository to list the raw findings of the work it publishes. Naturally, the vast majority of readers will not delve into these endless data charts and extensive explanations of the statistical analyses used. But the handful of colleagues and competitors who do so will provide an expert assessment of the data's quality. Ultimately, the entire scientific community will benefit from this strengthening of communalism, as suggested by a survey conducted by researchers at the Centre Cochrane at the Hôtel-Dieu Hospital in Paris.[4] Their study consisted of a comparison of the accuracy of information regarding drug trials, notably the occurrence of rare but serious events, in the ClinicalTrials.gov register and in a publication. They began by observing that half the trials posted on ClinicalTrials.gov had never been published, as has often been pointed out. But most importantly, they observed that in published studies, "the flow of participants [that is, the reasons for which a patient enters or leaves the research procedure], efficacy results, adverse events, and serious adverse events . . . are more completely reported at ClinicalTrials.gov than in the published article." In other words, medicine will make greater advances if researchers provide their raw data rather than publishing it in journals. This conclusion is certainly valid for other fields of research.

Down with the Impact Factor

This updating of the scientific publishing system will only bear fruit if it is matched with a deep reform of the way researchers are evaluated throughout their careers. Announced on May 13, 2013, the San Francisco Declaration on Research Assessment (DORA) called for radical reform by demanding that researchers no longer be assessed based on their impact factor (that is, the average number of citations of articles published in a journal over the last two years).[5] Behind this apparently technical question was a wide-scale initiative on the part of the international community of biologists to tackle the deteriorating quality of articles in their field.

The list of those who signed DORA constitutes an alliance as vast as it is unprecedented: thousands of researchers throughout the world, the editors of highly prestigious scientific journals (*Science, the Journal of Cell Biology, the EMBO Journal, Development*), and a multitude of professional societies that include DORA's initiator, the American Society for Cell Biology. In the words of one of the declaration's authors, its signatories called for "an insurrection" against using the impact factor to evaluate researchers, notably in hiring decisions and the awarding of funding. The first signatories have since been joined by thousands of others, as well as by hundreds of institutions, including, in France alone, the CNRS, Inserm, and the Pasteur Institute.

Criticism of journals' impact factor is not a new phenomenon. The reasons that impact factors cannot be considered a good indicator of the reach of a researcher's work have long been apparent. I'll mention five. Impact factors measure the impact of a journal, not of an article. They measure an average, which can easily be thrown off by one or two heavily cited articles—the publication of the historic article describing the sequencing of the human genome in *Nature* in 2000, which has been cited more than ten thousand times, allowed the British journal to achieve and retain the highest impact factor. Impact factors cannot measure the lasting influence of published research, since they are calculated based on the previous two years. They do not take into account whether articles were cited favorably or whether they were cited in order to question their data. Finally, impact factors can easily be manipulated by editors, notably by favoring "hot articles" dealing with the most controversial fields (stem cells, GMOs, global warming, and so on), which serve as guarantees of abundant citations.

Bibliometric indicators, which have the best-known impact factor, were originally conceived between the world wars to facilitate the work of librarians deciding which publications to subscribe to. They acquired a new function in the 1990s when public research funders faced with an economic slowdown started to evaluate scientific institutions' productivity and effectiveness. In the 2000s this pressure to evaluate through bibliometrics shifted from the collective to the individual, or, in other words, from the laboratory to the researcher.

This explains why most funding applications and evaluation forms now ask researchers to include their h-index, which is named after the last initial of its inventor, the American physicist Jorge E. Hirsch, and is defined as n number of articles by an author that have received at least n citations. The sociologist of science Yves Gingras deplores the epidemic of h-indices that has hit the scientific community, given that this index "has none of the basic properties necessary for a good indicator. The index is poorly constructed and even dangerous when it is used as an aid to making decisions, for it can generate pernicious effects."[6] These would include hiring a researcher who has published ten articles each cited ten times ($h = 10$), rather than the author of three articles each cited one thousand times ($h = 3$), despite the fact that the latter author's reputation is obviously far greater.

Intrusive Bibliometrics

Equally absurd is the spread of bibliometric criteria supposed to make it possible to evaluate the reputation of journals, institutions (particularly academic ones), and even researchers. Gingras offers an amusing demonstration of the absurdity of the overly influential Shanghai ranking of top universities: "Since French universities are all directly under the authority of the Ministry of Higher Education, it would be easy to quickly move to the top of the Shanghai ranking; one would just have to apply Napoleon's solution by creating a single official university, the University of France. With all the publications gathered under this designation, France would be certain to be ranked first."

But there remains a gnawing question, to which Gingras admits he does not have an answer. How did a scientific community whose cardinal virtues are supposed to be critical analysis and rationality collectively allow itself to be evaluated by such clearly unfounded bibliometric criteria? It is stunning to consider the extent to which researchers for whom Merton's "organized skepticism" is a norm have accepted the tyranny of bibliometric indicators without delving into the way they are calculated. Or, more generally, the fact that they have allowed these indicators to become the criteria by which both their careers and their reputations are evaluated. This situation is better illustrated by an anecdote than a long theoretical gloss. At the beginning of a symposium held on a beautiful Mediterranean island with limited public transportation, a dozen biologists were waiting at the airport to get to the conference venue. A single taxi appeared. Who would get in first? One of the researchers suggested they base the decision on their h-indices. According to a witness, she was not kidding.

Yet everyone knows that publishing in a journal with a high impact factor guarantees neither the quality of the research nor its originality. To consider only the most famous journals, *Nature* and *Science* have a commercial strategy equally aimed at the scientific community and at the general audience and decision makers in the sciences, via their articles' guaranteed media impact. Fashions come and go, but they always play a dominant role in the choice of whether to have a study evaluated. In the late 1990s the British journal *Nature* was known for taking a benevolent approach to any study tending to show the dangers of GMOs. And in the 2000s, while the Bush administration bowed to pressure from the religious Right and considered tightening regulations on stem cell research, the American journal *Science* was specifically seeking articles showing the benefits to be expected from embryonic stem cells (see Chapter 2).

While researchers are perfectly aware of all this, publishing an article in one of these prominent journals is all it takes for a scientific career to suddenly go into overdrive. Since they often lack the time to seriously study a candidate's qualifications or submitted project, researchers—who double as their colleagues' evaluators—frequently succumb to the temptation to rely on the impact factor of the journals in which candidates publish when evaluating their work. Many researchers will privately admit to this. The French Academy of Sciences observed, "It must be recognized that endorsement by publication in a journal with a high impact factor is supported by a not insignificant part of the scientific community (principally in biology) as well as by certain research organizations in France that use journals' impact factors as a criterion for selecting researchers. Many researchers defend this system, because it is a very simple and very quick method of obtaining an evaluation without expending much effort. But one must be aware that applying automatic methods always leads to grave injustice."[7]

Publish Less, Publish Better

This dictatorship of the impact factor is only one aspect of a trend affecting all sectors of the economy, both public and private, which is the reliance on allegedly objective numeric indicators to evaluate work. The harmful effects of the phenomenon are well known: competition between employees, deterioration of professional relationships within a company, workplace suffering sometimes leading to suicide.[8] Equally well known is employees' ambiguous resistance to being made to compete with each other: employees more often contest the manner in which the quantitative indicators are constructed than their actual principle, giving in to the temptation to show that they "outperform" their colleagues. As the psychoanalyst Bénédicte Vidaillet of the University of Lille-1 observes, "The ideology of evaluation has penetrated deep into employees' minds: while they massively criticize its effects, for reasons as significant as they are varied, they nonetheless cannot let go of the belief that more evaluation would resolve all the problems diagnosed. They want to be evaluated and rewarded individually, no matter the cost."[9]

This observation is absolutely accurate regarding the world of scientific research. Each laboratory has its lame duck, its layabout who "hasn't published in years," often the target of internal scorn and more or less silent mockery. Those laughing today prefer not to imagine the day when it will be their turn to be the lame duck or the layabout, whether because of the difficulty of the experiments they will undertake, the complexity of the questions they will try to elucidate, a lack of scientific creativity at a certain point in their lives, or experimental blunders in tackling a new subject.

Over the last few years, French scientific institutions, and particularly those connected to universities, have started hounding "non-publishers." Wouldn't they do better to be equally vigilant regarding "overpublishers" (see Chapter 6)? We have every reason to believe that those who publish so prolifically cannot attain these record-breaking levels of productivity without committing breaches of scientific integrity. Wouldn't it also be wise to consider how the major scientific publishers' marketing has powerfully contributed to turning bibliometric measurements into the alpha and omega of researchers' reputations? This highly profitable industry does not shy away from using its journals' impact factors for promotional purposes. It is not insignificant that the publishers of the journals with the highest impact factors, such as the British Macmillan Press *(Nature)* and the Dutch Elsevier *(Cell)*, turned down the invitation to sign DORA, unlike the learned societies that publish their own journals.

Putting an end to the dictatorship of impact factors and other bibliometric measures would certainly be a first step toward returning to a more qualitative than quantitative evaluation, in which scientific work's lasting influence would be appreciated over its fragile contribution to a fleeting trend, the collective nature of any scientific creation would be recognized, and it would be acknowledged that in the scientific community, reproducing a study is as important as being the first to publish it, since it proves its reliability.

Encouraging the publication of raw experimental data, as is now possible thanks to technological advances, and putting an end to the use of bibliometric measurements to evaluate researchers appear to be two immediate, practical, and conceivable measures to halt the explosion of scientific fraud. While they may seem dry or overly technical, they have the advantage of being realistic. But it would be hard for me to end this book without making a slightly more utopian suggestion: in order to regain its quality, science must slow down. The anthropologist Joël Candau made a similar argument in his manifesto for slow science:

Researchers, teacher-researchers, hurry up and slow down! Let's free ourselves from the Red Queen syndrome! Let's stop always wanting to run faster to ultimately run in circles or even backtrack! . . . Researching, thinking, reading, writing, and teaching take time. We don't have the time anymore, or less and less. Our institutions and, far beyond that, societal pressure promote a culture of immediacy, urgency, real time, constant tension, and projects that come one after another at an ever-faster rate. All this takes place not only at the expense of our lives—any colleague who isn't overworked, stressed out, and overbooked now seems like an apathetic or lazy eccentric—but also to the detriment of science. Fast science, like fast food, privileges quantity over quality.[10]

Faced with the malscience that so closely resembles the junk food served in fast-food restaurants, we need to slow down. And take our time. The time to think.

Singapore Statement on Research Integrity

The Singapore Statement on Research Integrity was developed in 2010 at the Second World Conference on Research Integrity.

Preamble

The value and benefits of research are vitally dependent on the integrity of research. While there can be and are national and disciplinary differences in the way research is organized and conducted, there are also principles and professional responsibilities that are fundamental to the integrity of research wherever it is undertaken.

Principles

- *Honesty* in all aspects of research.
- *Accountability* in the conduct of research.
- *Professional courtesy and fairness* in working with others.
- *Good stewardship* of research on behalf of others.

Responsibilities

1. *Integrity:* Researchers should take responsibility for the trustworthiness of their research.

2. *Adherence to Regulations:* Researchers should be aware of and adhere to regulations and policies related to research.

3. *Research Methods:* Researchers should employ appropriate research methods, base conclusions on critical analysis of the evidence and report findings and interpretations fully and objectively.

4. *Research Records:* Researchers should keep clear, accurate records of all research in ways that will allow verification and replication of their work by others.

5. *Research Findings:* Researchers should share data and findings openly and promptly, as soon as they have had an opportunity to establish priority and ownership claims.

6. *Authorship:* Researchers should take responsibility for their contributions to all publications, funding applications, reports and other representations of their research. Lists of authors should include all those and only those who meet applicable authorship criteria.

7. *Publication Acknowledgement:* Researchers should acknowledge in publications the names and roles of those who made significant contributions to the research, including writers, funders, sponsors, and others, but do not meet authorship criteria.

8. *Peer Review:* Researchers should provide fair, prompt and rigorous evaluations and respect confidentiality when reviewing others' work.

9. *Conflict of Interest:* Researchers should disclose financial and other conflicts of interest that could compromise the trustworthiness of their work in research proposals, publications and public communications as well as in all review activities.

10. *Public Communication:* Researchers should limit professional comments to their recognized expertise when engaged in public discussions about the application and importance of research findings and clearly distinguish professional comments from opinions based on personal views.

11. *Reporting Irresponsible Research Practices:* Researchers should report to the appropriate authorities any suspected research misconduct, including fabrication, falsification or plagiarism, and other irresponsible research practices that undermine the trustworthiness of research, such

as carelessness, improperly listing authors, failing to report conflicting data, or the use of misleading analytical methods.

12. *Responding to Irresponsible Research Practices:* Research institutions, as well as journals, professional organizations and agencies that have commitments to research, should have procedures for responding to allegations of misconduct and other irresponsible research practices and for protecting those who report such behavior in good faith. When misconduct or other irresponsible research practice is confirmed, appropriate actions should be taken promptly, including correcting the research record.

13. *Research Environments:* Research institutions should create and sustain environments that encourage integrity through education, clear policies, and reasonable standards for advancement, while fostering work environments that support research integrity.

14. *Societal Considerations:* Researchers and research institutions should recognize that they have an ethical obligation to weigh societal benefits against risks inherent in their work.

Source: World Conferences on Research Integrity, "Singapore Statement on Research Integrity," posted September 22, 2010, https://wcrif.org/guidance /singapore-statement.

Notes

1. Big Fraud, Little Lies

1. C. Babbage, *Reflections on the Decline of Science in England* (London: B. Fellowes and J. Boorh, 1830).
2. M. C. LaFollette, *Stealing into Print: Fraud, Plagiarism and Misconduct in Scientific P°ublishing* (Berkeley: University of California Press, 1992), 3.
3. A. Fagot-Largeault, "Petites et grandes fraudes scientifiques: Le poids de la compétition," in G. Fussman (ed.), *La Mondialisation de la recherche: Compétition, coopérations, restructurations* (Paris: Collège de France, 2011).
4. W. J. Broad and Nicholas Wade, *Betrayers of the Truth: Fraud and Deceit in the Halls of Science* (New York: Simon and Schuster, 1983), 60.
5. M. Grieneisen and M. Zhang, "A comprehensive survey of retracted articles from the scholarly literature," *PLoS One* 7, no. 10 (2012): e44118.
6. R. G. Steen, "Retractions in the scientific literature: Is the incidence of research fraud increasing?," *Journal of Medical Ethics* 37 (2011): 249–253.
7. F. C. Fang, R. G. Steen, and A. Casadevall, "Misconduct accounts for the majority of retracted scientific publications," *Proceedings of the National Academy of Sciences* 109, no. 42 (2012): 17028–17033.

8. D. Fanelli, "How many scientists fabricate and falsify research? A systematic review and meta-analysis of survey data," *PLoS One* 4, no. 5 (2009): e5738.

2. Serial Cheaters

1. Riken Research Paper Investigative Committee, "Report on STAP cell research paper investigation," Riken Institute, March 31, 2014, 9, http://www3.riken.jp/stap/e/f1document1.pdf.

2. Levelt Committee, Noort Committee, and Drenth Committee, "Flawed science: The fraudulent research practices of social psychologist Diederik Stapel," November 28, 2012, 53, https://www.tilburguniversity .edu/upload/3ff904d7-547b-40ae-85fe-bea38e05a34a_Final%20 report%20Flawed%20Science.pdf.

3. D. Stapel, *Faking Science: A True Story of Scientific Fraud*, trans. Nicholas J. L. Brown (self-pub., 2014), 206, https://slidelegend.com/faking -science-a-true-story-of-academic-fraud-wordpresscom_5a0e7ce61723 ddfde8b4285a.html.

4. E. S. Reich, *Plastic Fantastic: How the Biggest Fraud in Physics Shook the Scientific World* (New York: Palgrave Macmillan, 2009).

5. "How the biggest fraud in physics shook the scientific world," review of *Plastic Fantastic,* by Eugenie Samuel Reich, *ScienceBlogs,* August 31, 2009, https://scienceblogs.com/ethicsandscience/2009/08/31/book-review -plastic-fantastic.

6. Reich, *Plastic Fantastic,* 115.

7. "Rapport des sections 01 et 02 du Comité national du CNRS sur deux thèses doctorat," November 2003, 5, http://pratclif.com/universe /bogdanof/rap_bogda_2_ok.pdf.

8. P. K. Kranke et al., "Reported data on granisetron and postoperative nausea and vomiting by Fujii *et al.* are incredibly nice!," *Anesthesia and Analgesia* 90 (2000): 1004–1007.

9. P. K. Kranke et al., "The influence of a dominating centre on a quantitative systematic review of granisetron for preventing postoperative nausea and vomiting," *Acta Anesthesiologica Scandanavica* 45 (2001): 659–670.

10. M. R. Tramèr, "The Fujii story—a chronicle of naïve disbelief," *European Journal of Anaesthesiology* 30 (2013): 195–198.

3. Storytelling and Beautification

1. P. Medawar, "Is the scientific paper a fraud?," in David Edge (ed.), *Experiment: A Series of Scientific Case Histories First Broadcast in the BBC Third Programme* (London: BBC, 1964), 7–13.
2. F. Jacob, *Of Flies, Mice, and Men,* trans. Giselle Weiss (Cambridge: Harvard University Press, 1999), 127.
3. Jacob, *Of Flies, Mice, and Men,* 128.
4. A. Franklin et al., *Ending the Mendel-Fisher Controversy* (Pittsburgh: University of Pittsburgh Press, 2008).
5. M. Richardson, "There is no highly conserved embryonic stage in the vertebrates: Implications for current theories of evolution and development," *Anatomy and Embryology* 196 (1997): 91–106.
6. G. Holton, "Subelectrons, presuppositions and the Millikan-Ehrenhaft dispute," *Historical Studies in the Physical Sciences* 9 (1978): 161–224.
7. R. Feynman, *Surely You're Joking, Mr. Feynman!* (New York: Norton, 2010), 342.
8. G. Anichini, "Quand c'est la science qui bricole c'est du sérieux," in S. Boulay and M. L. Gérard (eds.), "Vivre le sable! Corps matière et sociétés," special issue, *Techniques and Culture* 61 (2014): 212–235.
9. M. Rossmer, "On scientific integrity, making research data publicly available and routes to open access," interview by D. Crotty, *Scholarly Kitchen* (blog), July 11, 2013, https://scholarlykitchen.sspnet.org/2013/07/11/interview-with-mike-rossner-on-scientific-integrity-making-research-data-publicly-available-and-routes-to-open-access/.
10. R. Van Noorden, "The image detective who roots out manuscript flaws," *Nature News,* June 12, 2015, https://doi.org/10.1038/nature.2015.17749.
11. "Beautification and fraud," *Nature Cell Biology* 8 (2006): 101–102.
12. A. B. Smith III, "Data integrity," *Organic Letters* 15 (2013): 2893–2894.
13. "Alleged image fraud by Shigeaki Kato lab at the University of Tokyo (alleged research misconduct)," January 24, 2012, YouTube video, 5:53, http://www.youtube.com/watch?v=FXaOqwanWnU.

4. Researching for Results

1. E. Masicampo and D. R. Lalande, "A peculiar prevalence of p values just below .05," *Quarterly Journal of Experimental Psychology* 65 (2012): 2271–2279.

2. E. P. LeBel et al., "PsychDisclosure.org: Grassroot support for reforming reporting standards in psychology," *Perspectives in Psychological Sciences* 8 (2013): 424–432.

3. N. C. Leggett et al., "The life of *p:* 'Just significant' results are on the rise," *Quarterly Journal of Experimental Psychology* 66 (2013): 2303–2309.

4. M. Pautasso, "Worsening file-drawer problem in the abstracts of natural, medical and social science databases," *Scientometrics* 95 (2010): 193–202.

5. D. Fanelli, "Negative results are disappearing from most disciplines and countries," *Scientometrics* 90 (2012): 891–904.

6. M. D. Jennions and A. P. Moller, "A survey of the statistical power of research in behavioral ecology and animal behavior," *Behavioral Ecology* 14 (2003): 438–445; J. E. Maddock and J. S. Rossi, "Statistical power of articles published in three psychology-related journals," *Health Psychology* 20 (2001): 76–78.

5. Corporate Cooking

1. C. D. Kelly, "Replicating empirical research in behavioral ecology: How and why it should be done but rarely ever is," *Quarterly Review of Biology* 81 (2006): 221–236.

2. H. Collins, *Gravity's Shadow: The Search for Gravitational Waves* (Chicago: University of Chicago Press, 2004).

3. F. Prinz, T. Schlange, and K. Asadullah, "Believe it or not: How much can we rely on published data on potential drug targets?," *Nature Reviews Drug Discovery* 10 (2011): 712–713.

4. C. G. Begley and L. M. Ellis, "Drug development: Raise standards for preclinical cancer research," *Nature* 483 (2012): 531–533.

5. Open Science Collaboration, "Estimating the reproducibility of psychological science," *Science* 349 (2015): aac4716.

6. N. J. Barrows et al., "Factors affecting reproducibility between genome-scale siRNA-based screens," *Journal of Biomolecular Screens* 15 (2010): 735–747.

7. P. Hert and G. Molinatti, "Attempt to settle a scientific controversy at a laboratory: Reputations, trust and loyalty" (unpublished manuscript).

8. A. Abbott, "Image search triggers Italian police probe," *Nature* 504 (2013): 18–19.

9. L. Zhang et al., "Exogenous plant MIR168a specifically targets mammalian LDLRAP1: Evidence of cross-kingdom regulation by micro-RNA," *Cell Research* 22 (2012): 273–274.
10. "Receptive to replication," *Nature Biotechnology* 31 (2013): 943.
11. L. Osherovich, "Hedging against academic risk," *SciBX* 4 (2011): 416.

6. Skewed Competition

1. D. Fanelli, "Why growing retractions are (mostly) a good sign," *PLoS Medicine* 10 (2013): e1001563.
2. L. Mucchielli, *Violences et insécurité: Fantasmes et réalité dans le débat français* (Paris: La Découverte, 2001).
3. A. M. Stern et al., "Research: Financial costs and personal consequences of research misconducts resulting in retracted publications," *eLife* 3 (2013): e02956.
4. P. Fronczak, A. Fronczak, and J. A. Hølyst, "Analysis of scientific productivity using maximum entropy principle and fluctuation-dissipation theorem," *Physical Review E* 75 (2007): 026103.
5. S. Kyvik, "Changing trends in publishing behavior among university faculty, 1980–2000," *Scientometrics* 58 (2003): 35–48.
6. L. Butler, "Using a balanced approach to bibliometric quantitative performance measures in the Australian Research Quality Framework," *Ethics in Science and Environmental Policy* 8 (2008): 83–92.
7. E. Wager, S. Singhvi, and S. Kleinert, "Too much of a good thing? An observational study of prolific authors," *PeerJ* 3 (2015): e1154.
8. K. W. Boyack et al., "A list of highly influential biomedical researchers, 1996–2011," *European Journal of Clinical Investigation* 43 (2013): 1339–1365.
9. C. Mary, "Sound and fury in the microbiology lab," *Science* 335 (2012): 1033–1035.
10. D. Raoult, "Sound and fury clarified," *Science* 336 (2012): 152–153.

7. Stealing Authorship

1. M. C. LaFollette, *Stealing into Print: Fraud, Plagiarism, and Misconducts in Scientific Publishing* (Berkeley: University of California Press, 1992).
2. W. S. Choi et al., "Duplicate publication of articles used in meta-analysis in Korea," *Springerplus* 3 (2014): 182.

3. M. Dansinger, "Dear plagiarist: A letter to a peer reviewer who stole and published our manuscript as his own," *Annals of Internal Medicine,* https://annals.org/aim/fullarticle/2592773/dear-plagiarist-letter -peer-reviewer-who-stole-published-our-manuscript.

4. M. Hvistendahl, "China's publication bazaar," *Science* 342 (2013): 1035–1939.

5. Hvistendahl.

6. Committee on Publication Ethics (COPE), "COPE statement on inappropriate manipulation of peer review processes," December 19, 2014, https://publicationethics.org/news/cope-statement-inappropriate -manipulation-peer-review-processes.

7. W. Yang, "Research integrity in China," *Science* 342 (2013): 1019.

8. "SCIgen—An automatic CS paper generator," accessed February 14, 2019, http://pdos.csail.mit.edu/scigen.

9. E. Seligman, *Math Mutation Classics: Exploring Interesting, Fun and Weird Corners of Mathematics* (New York: Apress, 2016), 90.

10. C. Labbé and D. Labbé, "Duplicate and false publications in the scientific literature: How many SCIgen papers in computer science?," *Scientometrics* 94 (2013): 379–396.

11. Nate Eldredge, *That's Mathematics!* (blog), http://thatsmathematics .com.

8. The Funding Effect

1. J. Lexchin, "Pharmaceutical industry sponsorship and research outcome and quality: Systematic view," *British Medical Journal* 326 (2003): 1167–1170.

2. J. E. Bekelman, "Scope and impact of financial conflicts of interest in biomedical research: A systematic review," *Journal of the American Medical Association* 289 (2003): 454–465.

3. J. K. Cataldo et al., "Cigarette smoking is a risk factor for Alzheimer's disease: An analysis controlling for tobacco industry affiliation," *Journal of Alzheimer's Disease* 19 (2010): 465–480.

4. R. N. Proctor, *Golden Holocaust: Origins of the Cigarette Catastrophe and the Case for Abolition* (Berkeley: University of California Press, 2012).

5. S. Foucart, *La Fabrique du mensonge: Comment les industriels manipulent la science et nous mettent en danger* (Paris: Denoël, 2013).

6. N. H. Harrit et al., "Active thermitic discovered in dust from the 9/11 World Trade Center catastrophe," *Open Chemical Physics Journal* 2 (2009): 7–31.

7. J. Quirant, *La Farce enjôleuse du 11 septembre* (Books on Demand, 2011).

8. G. E. Séralini, "Republished study: Long-term toxicity of a Roundup herbicide and a Roundup-tolerant genetically modified maize," *Environmental Science Europe* 26 (2014): 14.

9. M. de Pracontal, "OGM: Le labo secret de Séralini," Médiapart, October 13, 2012, https://www.mediapart.fr/journal/france/121012/ogm-le -labo-secret-de-seralini?onglet=full.

10. G.-E. Séralini, *Tous cobayes: OGM, pesticides, produits chimiques* (Paris: Flammarion, 2012).

9. There Is No Profile

1. D. S. Kornfeld, "Research misconduct: The search for a remedy," *Academic Medicine* 87 (2012): 877–882.

2. M. Bungener, "Les petits arrangements avec l'intégrité scientifique sont préoccupants," interview by Nicolas Chevassus-au-Louis, *La Recherche* 482 (2013): 70–74.

3. F. C. Fang, J. W. Bennett, and A. Casadevall, "Males are overrepresented among life science researchers committing scientific misconduct," *mBio* 4 (2013): e00640-12.

10. Toxic Literature

1. A. M. Stern et al., "Research: Financial costs and personal consequences of research misconduct resulting in retracted publications," *eLife* 3 (2014): e02956.

2. R. G. Steen, A. Casadevall, and F. C. Fang, "Why has the number of scientific retractions increased?," *PLoS One* 8 (2013): e68397.

3. A. M. Michalek et al., "The costs and underappreciated consequences of research misconducts: A case study," *PLoS Medicine* 7, no. 8 (2010): e1000318.

4. J. Arrowsmith, "Phase II failures, 2008–2010," *Nature Review Drug Discovery* 10 (2011): 328–329.

5. P. Azoulay et al., "Retractions" (NBER Working Paper 18499, 2012).

6. M. Binswanger, "How nonsense became excellence: Forcing professors to publish," in I. M. Welpe et al. (ed.), *Incentives and Performance* (Cham, Switzerland: Springer, 2015): 19.

7. M. P. Pfeifer, L. Gwendolyn, and G. L. Snodgrass, "The continued use of retracted, invalid scientific literature," *Journal of the American Medical Association* 263 (1990): 1420–1423.

8. R. Trivers, B. G. Palestis, and D. Zaatari, *The Anatomy of a Fraud: Symmetry and Dance* (Antioch, CA: TPZ, 2009).

11. Clinical Trials

1. R. G. Steen, "Retractions in the medical literature: How many patients are put at risk by flawed research?," *Journal of Medical Ethics* 37 (2011): 688–692.

2. J. C. Camp et al., "Retracted publications in the drug literature," *Pharmacotherapy* 32 (2012): 586–595.

3. J. Couzin and K. Unger, "Cleaning up the paper trail," *Science* 312 (2006): 38–43.

4. P. F. White, H. Kehler, and S. Liu, "Perioperative analgesia: What do we still know?," *Anesthesia and Analgesia* 108 (2009): 1364–1367; L. Ø. Andersen et al., "Subacute pain and function after fast-track hip and knee arthroplasty," *Anaesthesia* 64 (2009): 508–513.

5. R. Zaryschanski et al., "Association of hydroxyethyl starch administration with mortality and acute kidney injury in critically ill patients requiring volume resuscitation: A systematic review and meta-analysis," *Journal of American Medical Association* 309 (2013): 678–688.

6. J. Wise, "Boldt: The great pretender," *British Medical Journal* 346 (2013): 18.

7. B. Deer, "How the case against MMR vaccine was fixed," *British Medical Journal* 342 (2011): 77–82.

12. The Jungle of Journal Publishing

1. H. Maisonneuve, "L'accès libre à la science va devenir la norme," interview by Nicolas Chevassus-au-Louis, *La Recherche* 491 (2014): 76–80.

2. C. Shen and B.-C. Björk, "'Predatory' open access: A longitudinal study of article volumes and market characteristics," *BMC Medicine* 13 (2015): 230.

3. R. Virkar-Yates, "Are predatory publishers an American export?," Se-mantico.com, July 26, 2013, http://www.semantico.com/2013/07/are-predatory-publishers-an-american-export-part-2/ (site discontinued).

4. J. Bohannon, "Who's afraid of peer review?," *Science* 342 (2013): 60–65.

13. Beyond Denial

1. W. J. Broad and N. Wade, *Betrayers of the Truth: Fraud and Deceit in the Halls of Science* (New York: Simon and Schuster, 1983), 11–12.

2. D. J. Kevles, *The Baltimore Case: A Trial of Politics, Science, and Char-acter* (New York: W.W. Norton & Company, 1998), 106.

3. Kevles, 138.

4. B. K. Redman and J. F. Merz, "Scientific misconducts: Do the punish-ments fit the crime?," *Science* 321 (2008): 775.

5. P. Mongeon, "Les Rétractions et leurs conséquences sur la carrière des coauteurs: Analyse bibliométrique des fraudes et des erreurs dans le do-maine biomédical" (master's thesis, University of Montreal, 2013).

6. Council of the European Union, "Council Conclusions on Research Integrity," adopted December 1, 2015, http://data.consilium.europa.eu/doc/document/ST-14853-2015-INIT/en/pdf.

7. S. L. Titus, J. A. Wells, and L. J. Rhoades, "Repairing research integ-rity," *Nature* 453 (2008): 980–983.

8. Titus, Wells, and Rhoades.

9. G. Koocher and P. Keith-Spiegel, "Peers nip misconduct in the bud," *Nature* 466 (2010): 438–440.

10. L. Letellier, "Osons parler de la fraude scientifique," *Le Journal du CNRS*, no. 378 (2014): 26–27.

11. S. L. Titus and X. Bosch, "Tie funding to research integrity," *Nature* 466 (2010): 436–437.

14. Scientific Crime

1. P. Le Hir, "Le gène de l'obésité à l'épreuve du soupçon," *Le Monde*, April 21, 1998.

2. J. P. Brun and J. Lenfant, "Pourquoi taire une fraude scientifique?," *Le Monde*, July 18, 1998.

3. Collective, "L'intégrité, une exigence de la recherche," *Le Monde*, September 15, 2014.

4. J. Ninio, "Doubts about quantal analysis," *Journal of Neurophysiology* 98 (2007): 1827–1835.

5. R. K. Merton, *The Sociology of Sciences: Theoretical and Empirical Investigations* (Chicago: University of Chicago Press, 1973), 204.

6. Merton, 323.

7. Merton, 311.

8. R. K. Merton, *Social Theory and Social Structure* (New York: Free Press, 1968), 195.

9. R. Nuwer, "Lawless labs no more," *New Scientist* 227 (2014): 27–32.

10. D. J. Kevles, *The Baltimore Case: A Trial of Politics, Science, and Character* (New York: Norton, 1998).

11. D. Baltimore, "Baltimore's travels continued," *Issues in Science and Technology* 19, no. 4 (2013), http://issues.org/19-4/baltimore.

15. Slow Science

1. F. S. Collin and L. A. Tabak, "Policy: NIH plans to enhance reproducibility," *Nature* 505 (2014): 612–613.

2. A. Meadows, "To share or not to share? That is the (research data) question . . . ," *Scholarly Kitchen* (blog), November 11, 2014, https://scholarlykitchen.sspnet.org/2014/11/11/to-share-or-not-to-share-that-is-the-research-data-question/.

3. P. Doshi et al., "Restoring invisible and abandoned trials: A call for people to publish the findings," *British Medical Journal* 346 (2013): f2865.

4. C. Riveros et al., "Timing and completion of trial results posted at ClinicalTrials.gov and published in journals," *PLoS Med* 10, no. 12 (2013): e100566.

5. DORA—San Francisco Declaration on Research Assessment (website), accessed February 15, 2019, https://sfdora.org.

6. Y. Gingras, *Les Dérives de l'évaluation de la recherche: Du bon usage de la bibliométrie* (Paris: Raisons d'agir, 2014).

7. Académie des sciences, *Les Nouveaux Enjeux de l'édition scientifique* (Paris: Académie des sciences, 2014), 26.

8. C. Dejours, *L'Évolution du travail à l'épreuve du réel: Critique des fondements de l'évaluation* (Paris: INRA Éditions, 2003).

9. B. Vidaillet, *Évaluez-moi! Évaluation au travail: Les ressorts d'une fascination* (Paris: Éditions du Seuil, 2013).

10. J. Candau, "Pour un mouvement Slow Science," Université Nice Sophia Antipolis, July 17, 2011, http://www.unice.fr/LASMIC/PDF/Appel%20 pour%20la%20fondation%20du%20mouvement%20Slow%20 Science.pdf (site discontinued). Available at http://archive.fo/r2mMq.

Acknowledgments

The author would like to warmly thank Giulia Anichini, Boris Chenaud, Sophie Gaudriault, Grégoire Molinatti, Yves Sciama, and Hélène Staes, whose comments on an early draft of this book led to significant improvements. He would also like to thank all the researchers, notably Jean-Pierre Alix, Samuel Alizon, Claude Bagnis, Jeffrey Beall, Laurent Boiteau, Enrico Bucci, Martine Bungener, Antoine Danchin, Pascal Degiovanni, Pierre-Henri Duée, Marc Fellous, Alexis Gautreau, Yves Gingras, Michèle Hadchouel, Philippe Jarne, Cyril Labbé, Michèle Leduc, Hervé Maisonneuve, Jacques Ninio, Marie-Paule Pileni, Livio Riboli-Sasco, Claire Ribrault, François Rougeon, Benoît Vingert, Francis-André Wollman, and the late Nicole Zsurger, who agreed to discuss scientific fraud with him, providing previously unavailable information and pertinent remarks.

Index

Note: Page numbers in *italics* indicate figures.